m
media
MANUAL

nd

Sound Techniques for Video and TV

m
media
MANUAL

Sound Techniques for Video and TV

Second Edition

Glyn Alkin

Focal Press
An imprint of Butterworth-Heinemann Ltd
Linacre House, Jordan Hill, Oxford OX2 8DP

⟨R⟩ A member of the Reed Elsevier plc group

OXFORD   LONDON   BOSTON
MUNICH   NEW DELHI   SINGAPORE   SYDNEY
TOKYO   TORONTO   WELLINGTON

First published as *TV Sound Operations* 1975
Second edition 1989
Reprinted 1992, 1994, 1995

© Butterworth–Heinemann Ltd 1989

British Library Cataloguing in Publication Data
Alkin, Glyn
    Sound techniques for video and TV – 2nd ed.
    1. Sound recording and video recording
    Techniques. Manuals
    I. Title   II. Alkin, Glyn.   TV sound operations
    621.389'32

ISBN 0 240 51277 4

Library of Congress Cataloguing in Publication Data
Alkin, E. G. M.
    Sound techniques for video and TV/Glyn Alkin – 2nd ed.
    p.   cm.
    Rev. ed. of TV sound operations. 1975
    ISBN 0 240 51277 4
    1. Sound – recording and reproducing.   2. Television –
    production and direction.   3. Acoustical engineering.
    I. Alkin, E. G. M.   TV sound operations.   II. Title.
    TK7881.4.A42   1989
    791.45'024-dc 19   88-38287

Printed and bound in Great Britain at the University Press, Cambridge

Contents

Introduction

This book is intended for anyone who is involved in the reproduction of sound with vision, from the enthusiastic amateur, starting to shoot with a video camera (and about to find out that it is usually easier to get good pictures than good sound), to the professional television or film sound operator with the latest equipment for dealing with the most complex productions, including stereo.

Methods of handling each type of production situation are dealt with in detail, care being taken throughout to explain the underlying motivation and principles involved. The book should therefore be of interest to professional and amateur video and movie makers, users of audio-visual aids and anyone interested in the reproduction of sound.

The book includes sufficient technical information to enable the operator to understand the basic elements of the equipment and get the best out of it, audio theory being treated in a simple, non-mathematical manner.

The book contains a great deal of information, and the format, with a complete topic on each page, makes it easy for the reader to pick out subjects of particular interest, or to use it for reference.

Acknowledgment
The author wishes to acknowledge the assistance of Adrian Bishop-Laggett, President of the Institute of Broadcast Sound, in collecting information and checking the draft. As Senior Sound Supervisor in BBC television Adrian is at the forefront of the latest developments in television sound operations.

Sound in a Visual Medium

If you try the simple experiment of watching television for any length of time with the sound turned off three important facts will become obvious.
1. Most of the information and substance of practically all productions are contained in the sound.
2. Most pictures tend to lose realism and impact when you cannot hear the accompanying sound.
3. It is difficult to concentrate, even on the most visual material, unless you can hear sounds that are at least in some way related to the picture. The old idea of the piano accompaniment to a silent motion picture was a crude attempt to underline the emotional message of the picture and assist concentration by occupying the aural sense and covering the noises in the auditorium.

Relative importance of picture and sound
When sound and picture are combined they are both equally important and should complement each other. To say that either can stand up on its own is not necessarily to praise it. Rather it suggests that the two media are not properly integrated.

It is all too easy to assume that the presence of the picture makes the achievement of good sound production less critical. This is not true. If the sound is unclear or incomplete the picture will not help. It will merely draw the attention of the viewer to what he is missing.

Relative complexity of picture and sound
It is also easy, particularly when first starting to work with video, to assume that the audio aspect will be comparatively simple and will somehow look after itself. After all, if you can see a source of sound, apparently at close range, you expect to hear it easily. However, with long-focus lenses it is a simple matter to take close-up pictures from a considerable distance, but picking up the corresponding sound and separating it from the background noises could present an insuperable problem.

Why should this be? Why is it not just a simple matter of fixing a microphone on the camera and pointing them both in the same direction? After all, cameras are available now with microphones that 'zoom', i.e. alter their angle of acceptance, in conjunction with the zoom lens on the camera. This might appear to be the ideal way to solve the problem, but the fact is that, unless the programme material is very simple and the conditions ideal, the results are likely to be very disappointing and unprofessional.

So it is worthwhile to consider some of the basics of sight and sound and analyse some of the differences between the two media.

Vision

Our eyes provide us with two sorts of vision, central and peripheral, giving us a wide and narrow view simultaneously. Our central vision has a very narrow angle (less than 1°) which we can focus at varying distances. Around this we have peripheral vision of about 180° giving us an overall view in much softer focus. When we want a narrow view, or to see a distant object, we locate the spot using our peripheral vision and focus our central vision on to it. For a wide view we make more use of our peripheral vision but scan the scene to pick out details with our central vision. We do not do as the zoom lens does, change the whole perspective by changing the apparent distance of the object.

Sound

Unlike visual objects, which (unless luminescent) require to be illuminated before they can be seen, sound sources produce their own energy, which they radiate in all directions whether we wish to hear them or not. Our ears, in association with our brain, give us some degree of selectivity, but in general if a sound is loud enough it is inescapable and very difficult to confine.

It is true that pictures can be more compelling than sounds, and if the sound and picture conflict the viewer will tend to believe the picture, but he will be left with a sense of unreality and a (possibly unexplained) feeling that something is wrong, which can undermine his appreciation of the programme. But if the sound and picture are perfectly in accord the result can have more impact than simply the sum of the two.

Listening via a microphone and loudspeaker can be very different from hearing the same sound direct.

The Hearing Process

For many people their first attempt to use microphones is the first time that they come up against the problems of acoustics. They may make video recordings and find, on replay, that although the pictures reproduce more or less as expected, the sound is very different and possibly unusable, even though the equipment used is technically immaculate.

This can be largely due to the basic differences between listening in person and through a microphone. It is, perhaps, not generally realised that the combination of our ears and our brain provides us with an incredibly complicated mechanism that enables us to hear selectively.

The 'cocktail party' effect

If you are in the middle of a crowd of people, all talking to each other, it is usually possible to focus your hearing on a conversation that you find interesting, while effectively pushing all the other sounds into the background. You do not have to turn your head (although this helps) or move your ears like a cat. Your brain interprets the minute differences in time, volume and quality of the sound as it reaches each of your ears in turn. This gives you the ability to judge the direction of the sound sources and discriminate accordingly. On the other hand, microphones, although they may have directional characteristics, are not able to vary them from instant to instant. They pick up everything within range, including all the reflected sounds that bounce back from the walls, floor and ceiling, reproducing them in proportion. Moreover the volume range of the sound sources is usually greater than is acceptable in domestic listening conditions, and mere reduction could make the quieter elements inaudible.

In many ways sound is more awkward to control than vision. It can curve around obstacles that are smaller than its wavelength (and that includes many items of scenery and furniture) whereas light generally travels in straight lines and can easily be contained.

All this means that, while it is usually possible to predict what a camera will see in a given direction, it is seldom possible to assess what a microphone will pick up without taking into account many complex factors.

Sound waves

Sound waves have a large range of physical dimensions comparable to many items of furniture and equipment etc. Sound is capable of bending around obstacles whose dimensions are smaller than the wavelength.

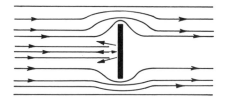

Control of light

Light travels in predictable straight lines. The picture from the camera is limited to the angle of view of the lens. The light which provides the source of reflected energy can often be controlled in intensity and confined in area.

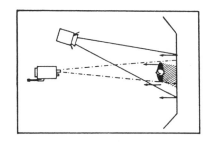

Control of sound

The sound source radiates over a wide angle and variable direction. The intensity is under the artist's control and the volume range often excessive and unpredictable. Sound reaches the microphone via direct and multiple reflective paths.

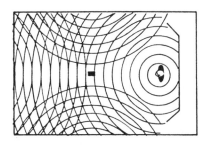

Unwanted sound

Noise or unwanted sound can reach the live face of the microphone via multiple reflective paths. Middle and low frequency sounds can curve around from behind sets or any other obstacle smaller than their wavelength.

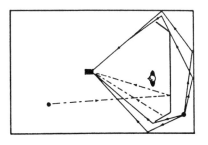

What is Sound?

Frequency

Sound is an aural sensation created by vibrations that have a frequency, i.e. repetition rate, within the range that we can hear. The lowest frequency we can hear is about 15 hertz (cycles of repetition per second) and the highest about 20 000 Hz, but our high-frequency sensitivity normally diminishes progressively from middle age.

The simplest forms of sound vibrations, into which all sounds no matter how complex can be resolved, are pure tones such as can be produced by tuning forks. Tuning forks produce simple harmonic motions which could be described by plotting the displacement of a swinging pendulum against time. Each cycle of vibration is indicated by an excursion each side of its rest position, resulting in the familiar sine wave curve.

Velocity

When sound travels through the air, which is the way we usually hear it (we can also hear by bone conduction), the vibrations consist of alternately increasing and decreasing air pressures, which travel outward from the source at a speed of around 340 metres per second (1120 feet per second). The actual velocity depends on the temperature of the air through which it passes, increasing with heat and, to a much lesser extent, humidity. Sound propagation through solids tends to be much more rapid and efficient than through air; for example, through steel it can travel about fifteen times as fast.

Wavelength

The velocity of sound is related to wavelength, i.e. the physical distance between one wavecrest and the corresponding position on the next one, in determining frequency (velocity = frequency × wavelength). This is why orchestral wind players have to keep tuning their instruments, lengthening the air column as the concert hall and their instruments warm up.

Phase

A sine wave can be represented by a vector which rotates through 360° for each complete cycle. The length of the vector represents the *amplitude* of the wave (maximum power), and the height above and below the base line represents the strength and polarity of the wave at that moment in time. When several waves are superimposed the resultant will be their vector sum or difference according to their relative *phase*, i.e. their displacement in time. If the phase angle between them is less than 90° they will add together, but if it is greater they will subtract. If two waves of equal frequency and 180° apart are combined they will cancel out completely. Phasing plays an important part in the design of microphones.

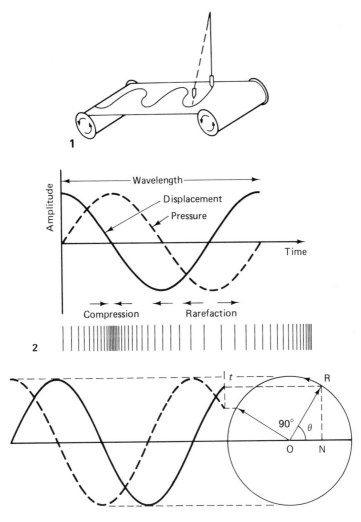

1. Simple harmonic motion. If the paper is pulled with a smooth motion past a pendulum swinging in one plane, the trace will be a simple harmonic (sine wave curve). The number of times the curve repeats itself per second is the *frequency*.

2. Sound is transmitted through the air by a series of alternating compressions and rarefactions, i.e. changes of air pressure above and below the prevailing atmospheric pressure. Maximum pressure variations occur during maximum *change* in displacement of the exciting force, i.e. when crossing its normal state. Pressure variations are minimum when the force is at a maximum.

3. Wave motion can be represented by rotating vectors. The radius OR represents the amplitude of the wave and the resultant value above or below the base line is RN. As the vector rotates plotting the angle θ against RN is equivalent to plotting θ against $\sin \theta$, hence the curve is called a sine wave. The diagram illustrates two sine waves with a relative phase angle of 90° at time t.

15

Sound Intensity and Loudness

Sound intensity
The intensity of a sound depends on the power of the source and the manner in which it has travelled. If the source is very small, i.e. a point source, the wave front radiates as an expanding sphere and the intensity (which depends on the area of the wavefront) diminishes as the square of the distance. If the source is large, or a long way away, the wavefront is practically flat (plane) and the intensity does not vary much with distance of travel. All this assumes that the sound is travelling freely in the open air, but in most cases it has to traverse rough surfaces where it loses energy by friction and absorption.

Loudness
When the sound reaches the listener his impression of its loudness depends not only on its intensity but also on many other factors, some of which are largely subjective and impossible to measure accurately. The factors include the environment of the listener and the ambient noise level, the pitch and character (harmonic structure) of the particular sound, the loudness of the sound that preceeds and follows it, and even such factors as the listener's appreciation of the programme material. Unwanted sound (noise) can sound louder than wanted sound of equal intensity. A discord can sound louder than a chord.

Just as our appreciation of variations in musical pitch is determined by the ratio, rather than the numerical value, of the frequency interval (e.g. two frequencies an octave apart are always a ratio of 2:1, whether at the low end of the scale where the frequency difference is comparatively small or at the high end where it is very much greater), so the difference in loudness depends on the ratio, rather than the numerical difference in intensities. In fact the ear has a logarithmic response; it equates equal changes of loudness with equal changes in the logarithm of the intensity.

For this reason relative intensities are expressed in bels (B), a unit that compares the logarithm of their powers to the base 10. In practice the bel is too large a unit and the decibel (dB) is used. This represents about the smallest discernible change in audio power, equivalent to about 26%.

$$\text{Decibels} = 10 \log \frac{\text{intensity 2}}{\text{intensity 1}}$$

Intensity is proportional to pressure squared, so

$$\text{Intensity} = 20 \log \frac{\text{pressure 2}}{\text{pressure 1}}$$

This makes it possible to express audio levels, gains and losses in simple figures instead of having to use ratios.

16

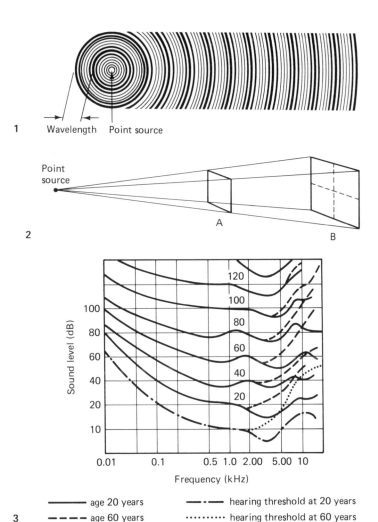

1 Wavelength   Point source

Point source

A

2

B

3

—————— age 20 years   — · — · — hearing threshold at 20 years

– – – – age 60 years   · · · · · · · · · · hearing threshold at 60 years

1. Sound radiates from a point source as a series of expanding spheres of alternating compression and rarefaction. At a distance, or if the source of sound is large, the wave front is practically flat.

2. With a point source radiating equally in all directions, if B is twice as distant as A the pressure at B is distributed over twice the area, i.e. the intensity is proportional to the square of the distance from the source.

3. Contours of equal loudness plotted against frequency for pure tones at various intensities. The various curves represent different values in *phons*. The phon is a unit that takes into account the unequal sensitivity of the ear. It relates the intensities of sounds at various frequencies to an equally loud sound at 1 kHz. The solid line represents average sensitivities at age 20 years, the dotted line at age 60. Note how our sensitivity to low frequency diminishes at low volume. This is the reason for tone-compensated loudness controls. The broken line at the bottom represents minimum levels of audibility.

17

Sound quality is subject to a number of different types of distortion which must be eliminated for high fidelity reproduction.

Sound Quality

To achieve high quality sound it is necessary to analyse the various characteristics that make for good sound and consider the possible distortions that can spoil it.

Frequency response
Ideally the overall response of the system should be level over the desired frequency range, typically 30 Hz to 15 kHz. At first sight (see table opposite), it would appear that the extreme high frequency range is unnecessary, especially as the main part of the musical scale and bulk of the audio power is in the middle/lower register. The high frequency range is important, however, as it contains the overtones, or upper partials, which give the various sound sources their characteristic timbre or quality. These overtones can extend to a very high order.

Harmonic structure
The difference between a particular note played on, for example, a piano or a violin depends upon the shape of the resulting waveform. Every waveform, no matter how complex, can be resolved into a fundamental (or lowest frequency), which determines the pitch of the note, and a series of overtones of higher frequency. The latter may be harmonics, i.e. multiples of the fundamental frequency, or unrelated inharmonics such as feature largely in the initial impact of percussion instruments. These starting transients are particularly significant in establishing the timbre, i.e. quality, of an instrument.

Amplitude response
Variations in level of the input signal should be matched throughout by equal variations in the reproduced output otherwise *amplitude distortion* is present. An example of the deliberate use of amplitude distortion occurs in compression amplifiers where the variations in input level result in lesser variations in the output.

Linearity
Any non-linearity that occurs in the processing of the waveform, e.g. overloading, causing flattening of the curve, can result in the production of spurious harmonics or *harmonic distortion*.

Intermodulation
If the various frequency components in a signal are caused to intermodulate with each other (multiply instead of add together) due to non-linearity in the system, *beat* frequencies (i.e. sum and difference tones) are produced which can form a long series of harmonics some of which will be discordant. This is known as *intermodulation distortion.* It creates an effect known as *acoustic roughness* which is found to be most objectionable when the beat rate (frequency difference) is of the order of 50 Hz for frequencies around 3 kHz.

1. The tonal range of voices and some musical instruments. a: Soprano. b: Contralto. c: Tenor. d: Baritone. e: Bass. f: Piccolo. g: Violin. h: Viola. i: Flute. j: Clarinet. k: Trumpet. l: Bass clarinet. m: Bassoon. n: Cello. o: Tuba. p: Double bass.
Only the fundamental frequencies are represented. The overtones extend throughout the upper audio frequency range.

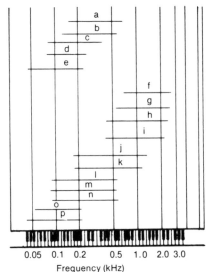

Frequency (kHz)

2. Analysis of a complex waveform into its fundamental and harmonic frequency components. The shape of the final waveform depends not only upon the number and strength of the individual overtones but also upon their relative phase, i.e. their time relationship to each other.

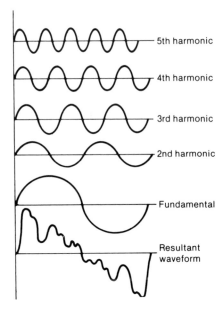

The Sound Director (Audio or Sound Supervisor) is the audio equivalent of the Vision Director.

The Function of Audio

The person in charge of sound is variously known as the Sound Director, Audio Supervisor or Sound Supervisor, and is commonly thought to be 'the person responsible for the quality of the sound'. This is true, but it is only part of his or her responsibility. Although certainly the custodian of technical quality, he or she is rather more concerned with the *character* of the sound. The main task is to interpret the writer's or composer's intentions.

Reproducing sound for domestic listening conditions is seldom a matter of producing a faithful copy of the original, because in most cases the listening conditions are so different. Some aspects have to be made larger than life and some smaller, with a delicacy of balance made more acute by restricted listening levels.

The volume range of even the simplest program material is usually too great for comfort in the living room, and once the audio level has been suitably reduced the balance between the various elements in the production becomes more critical.

The first task is to arrange suitable acoustic conditions to enhance the sound and promote the comfort and performance of the artists. The acoustic environment can have a profound effect on performance as well as the audio output.

Sound balance and control
Having produced the right conditions for the production of the sound the next task is to balance it. The term 'sound balance' means the positioning of the artists and microphones and the mixing of their outputs to achieve a balanced artistic effect within the confines of the medium. It involves the exercise of choice on behalf of the listener and is the aural equivalent of camera direction.

In addition to achieving balance between the sources from instant to instant the sources have to be controlled for three basic purposes.
1. To create emphasis — possibly even from a single source — by relating volume to importance.
2. To curb the original sound level variations to suit recording and broadcasting parameters and domestic listening conditions.
3. To control the *character* of the audio by adjusting the frequency response and selecting the degree of 'presence' and impact to suit the programme material.

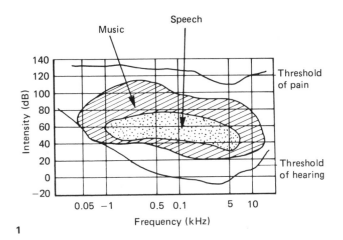

1

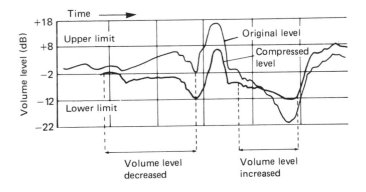

2

1. *Volume and frequency range.* An approximation of the volume and frequency range encountered in music and speech, heard at normal distances. In practice with close-working multi-microphone technique, or in some drama situations, the actual level variations at individual microphones can be much greater. Some modern pop groups surpass the threshold of pain!

2. *Anticipating peaks.* The process of sound control requires a sensitive appreciation of the programme material and is achieved by intelligent anticipation with the aid of a script or score. Here, volume variations of about 40 dB have been contained within a 20 dB limit without losing impact. The operator has realised the importance of giving the sudden peaks their full artistic value and has made way for them by creeping the level down slowly when it will be least noticeable.

The acoustic conditions can have an important bearing on the whole technique of sound production.

Acoustics

When a sound is made it produces waves of alternately compressed and rarefied air which travel out from the source at a speed of approximately 340 m/s (1120 ft/s). It is important to realise that the bulk of the air does not have to travel with the wave, which consists of air particles that merely move backwards and forwards about their median position, passing their energy from one to another rather like rail cars shunting.

When the sound waves meet an obstacle, such as the walls, floor or ceiling of a room, they are either reflected or absorbed to a degree depending upon the nature of the substance they encounter. If the material is dense, porous and of sufficient thickness, the sound waves will penetrate it and use up energy in finding their way through the particles, which are caused to vibrate and expend energy in the form of heat.

The effect of size

If the sound waves encounter a substance with a hard surface, and if it is larger than the wavelength of the sound, most of the energy will be reflected in much the same manner as a mirror reflects light. If the sound is in a room it will continue to bounce around between the reflective surfaces until its energy has been dissipated. These reflections will tend to overlap and merge together to reinforce the volume of the sound. If the room is large, with reflective walls, the reflections will arrive after a lapse of time and the overlaps would blur the clarity of speech but could do much to enhance legato music.

Reverberation

The prolongation of sound by overlapping reflections is called *reverberation*. It should not be confused with *echo* which, strictly speaking, is a single repetition of the sound, so spaced from the original as to be separately identifiable, i.e. delayed by more than about 60 ms.

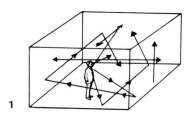

1

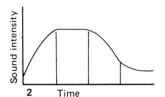

2   Time

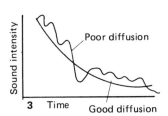

Poor diffusion

3   Time   Good diffusion

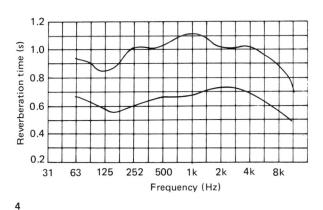

4

1. In an enclosure with reflective walls, sound bounces around in all directions between all the surfaces until it uses up all its energy.

2. When a continuous sound is produced in a reverberant room its intensity builds up as the reflections reinforce each other until a state of equilibrium is reached. When the source of sound is cut, the sound dies away slowly, the rate of decay depending on the reverberation time.

3. The effect that room acoustics have on a sound depends not only on the time the sound takes to die away but also on the smoothness of the decay. A smooth decay suggests good diffusion (well broken up surfaces).

4. Typical reverberation time/frequency curves for a large and a small television studio (upper and lower curves, respectively). They have about half the reverberation time of a typical radio studio of similar size.

23

Acoustic Effect, Reverberation

The most important features of the acoustics of a room or studio that affect the sound, and possibly the performance of the artists, are the reverberation time, the shape of the decay curve and the diffusion.

Reverberation time

The reverberation time is a measure of the time a sound takes to die away after the source providing it has stopped. Technically it is the time, in seconds, the sound takes to drop in volume through 60 dB (i.e. to one millionth of the original).

The reverberation time is related to the volume of the enclosure and the amount of absorbent material it contains. Rough guide figures are usually quoted for reverberation times around 1 kHz, but most rooms have reverberation times that vary considerably with frequency and this can give rise to marked colouration of the sound.

Typical mid-frequency reverberation times are:

Average living room	0.6 s	Small television studio	0.7 s
Light orchestral studio	0.8 s	Large television studio	1.0 s
Small monitoring room	0.2 s	Large concert hall	1–2 s

Decay curve

Almost as important as the reverberation is the shape of the decay curve, i.e. what happens to the sound immediately after the source is cut. A smooth decay curve suggests good diffusion.

Diffusion

Reverberation is caused by sound bouncing around inside a room until its energy has been converted into heat by friction on absorbent surfaces. For good acoustic conditions the reflective and absorbent surfaces should be well distributed so that focusing effects and standing waves are minimised.

Television studio acoustics

The problem about designing the acoustics of a television studio is that, unlike their counterpart in radio or recording; they cannot usually be allocated for one particular purpose. A large television studio may have to accommodate anything from a symphony concert (which sounds best with a reverberation time in excess of two seconds) to an open air sequence in drama which should have no reverberation at all. As it is possible to add artificial reverberation (see page 112) but not subtract it, television studios are usually made as dead as possible.

A dead acoustic also helps to damp down noise from all the movement of people and machinery inevitable in the television operation.

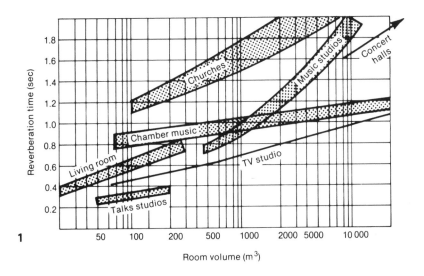

1

Room volume (m³)

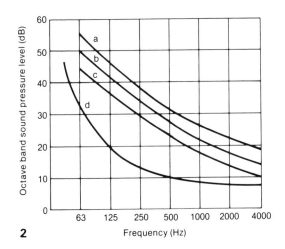

2

Frequency (Hz)

1. Recommended mid-frequency reverberation time for various sizes and functions of rooms.

2. Recording venues should be as quiet as possible. The table gives the accepted criteria for background noise levels in broadcasting studios: (a) BBC audience studios; (b) BBC television and sound studios (except drama); (c) BBC sound drama studios; (d) Quiet sound studios.

Our ears judge the distance of a sound by its relative volume and characteristic quality.

Co-ordinating Sound and Picture

In drama, more than any other type of production, there is an attempt to achieve a realistic effect. It is important to remember that sound can set the scene just as well as, and sometimes better than, vision. Listeners to radio drama, or anyone who has tried the simple expedient of listening to natural sounds with his eyes shut, are well aware of the ability of sound to conjure up atmosphere and a sense of environment without need of a picture. This must be equally true of television sound, whether the viewer realises it or not. Nor is it true only of televised drama. In practically every type of production a great deal of effort on the part of scene designers, lighting and camera directors goes into creating the illusion of three-dimensional reality on the screen. This can be underlined or undermined by good or bad sound.

Perspective
An important property of sound is its ability to suggest perspective, a sense of distance or third dimension, and this also must be in agreement with the visual perspective shown on the screen. Few things are as destructive to dramatic realism as to hear an artist appear to recede in sound as he approaches in vision, or vice versa. Such an effect can easily be produced accidentally, e.g. if an actor, approaching the camera, overruns the minimum retraction of the boom.

Distance
The ear tends to judge the distance of a sound source in the following ways:
1. By relative volume level. The ear has a poor memory for absolute level but is quick to notice a change. Obviously the louder the nearer.
2. Tonal characteristics. With the voice the sibilants and higher consonants are more pronounced at close range (hence the so-called 'presence filters' which increase output in the upper-middle frequency range).
3. Acoustic effect. More reverberation is associated with greater distance.
4. Frequency response. We recognise that in most real-life situations high frequencies are more affected by distance than low.
5. Time factor. The ear can detect the extent to which the direct sound precedes the reverberation, and thus can distinguish between a close sound in a reverberant room and a distant sound in a less reverberant one.

Perspective matching can be achieved by careful control of studio and set acoustics, microphone positioning, the addition of artificial reverberation and the subtle use of frequency correction.

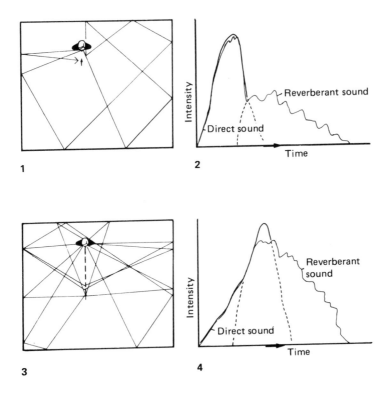

1. Artist working close to microphone. His direct sound reaches the microphone well in advance of the reverberation, which is at much lower volume.

2. The sort of decay curve to be expected from an artist making an impulsive sound at close range in a reverberant room. Our ears can recognise that the direct precedes the indirect sound, and this gives us the clue that the sound is close, but the room is reverberant.

3. Artist at some distance from the microphone. The reflections arrive at a similar time to the direct sound, which appears to be submerged in the reverberation.

4. Slower build-up and later decay of sound. Our ears associate this effect with distance.

Acoustic conditions are seldom ideal. It is essential to make the best of each situation.

Acoustic Treatment

When it is desired to pick up sound in premises which have not been designed for the purpose it may be possible to modify the acoustics without incurring too much expense.

Simple tests

First it is necessary to assess the characteristics of the room. Ideally this should be done by proper acoustic measurement, but a simple assessment can be made by making a loud impulsive noise (hand clap or drum rim-shot) and noting how long the sound takes to die away. Note particularly if there is any ringing or flutter, which can occur between reflective surfaces that are parallel to each other. Note also if the area is oppressive to speak in (too much high frequency absorption) or if the sibilants appear to flutter from the walls (too little high frequency absorption). If possible play an instrument, or music through a good loudspeaker, and note if the quality appears to vary as you walk about the room.

If you are not able to modify the acoustics, try to find a position in the room (probably assymetrical in relation to the room dimensions) where the effects are minimised, to place the action. It may be possible to reduce flutter and colouration by interposing screens or drapes between parallel walls and between the walls and the microphone.

Modifying acoustics

If you can make alterations the first thing to consider is to fit thick carpet and underfelt. Next is the provision of heavy drapes over the windows, as these tend to be very reflective and allow the ingress of noise. Normal curtain material will affect only the high frequency sounds. To reduce lower frequency reflections it is necessary to use heavy drapes, 10–15 cm (4–6 in) away from the windows.

Wide-band absorbers can be constructed comparatively simply (see diagram) and screens made to improve isolation and break up standing waves.

Acoustics can be adjusted by panel absorption units. The most useful type is the wide-band absorber that combines porous absorption for high frequencies with membrane absorption for the bass. Construction of a wide-band absorber: (a) perforated hardboard; (b) bituminous roofing felt, bonded to the back of the hardboard; (c) rockwool or glass fibre; (d) timber frame; (e) wall. The rockwool or glass fibre absorbs the high frequency sounds which penetrate the holes and the roofing felt. The roofing felt damps the hardboard due to its inherent sluggishness and causes it to act as a damped membrane absorber. A wide range of absorptions can be arranged by making up boxes of differing sizes and thicknesses.

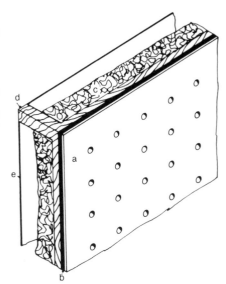

Typical absorption coefficients of wide-band absorbers. The high frequency absorption can be determined by the ratio of perforation of the hardboard. The holes act as cavity resonators (Helmholtz resonators) with the air space behind, for (a) 25%, (b) 5% and (c) 0.5% perforation.

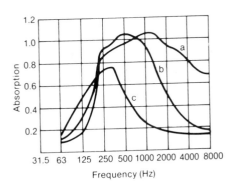

Acoustic Effect of Scenery

No matter how much care has gone into the design of a television studio, its acoustics are likely to be considerably modified when the scenery is brought into it. In fact where the action, as so often, takes place in a small parallel-sided set within a very large studio the acoustics of the set have more effect on the sound than those of the studio. Any reflections from the studio that reach the microphone are likely to be too delayed to be of positive use and should be eliminated as much as possible.

Acoustic planning
When planning a production, the sound supervisor should have a clear idea of the effect that the settings are likely to produce, so as to be able to offer advice to the scene designer.

Design considerations
The effect that the settings will have on the sound depends on:
1. The extent to which the material used reflects or absorbs sound.
2. The shape and size of each piece and thet angle it makes to the sound source and to the microphone.
3. The possibility of resonance of the material or within any cavities in the framework, e.g. rostra.

   This is particularly important in the case of music, where sympathetic resonances in the surrounding scenery or rostra can cause unpleasant colouration and spoil the tone of the instruments. Particular care must be taken in the design of rostra that come into direct contact with the instruments, e.g. those for celli, basses and percussion.

Acoustic effect of materials
In general, materials can be divided into the hard and the soft variety. Hard surfaces such as plywood, glass fibre, reinforced resin and aluminium tend to reflect the higher frequencies strongly and, if in fairly large panels, will absorb some bass by acting as a dampened membrane. Curtains, drapes, carpet and soft furnishings are very absorbent at high frequencies and can absorb the bass as well if spaced away from a hard surface to form a cavity.

The importance of shape
The shape of the settings and objects is very important. Look out for focusing effects. It is no good using the directional properties of a microphone to discriminate against noise in front of a reflecting surface that might actually focus it back on to the live face.

30

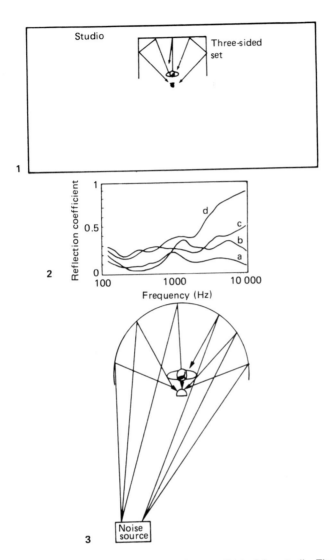

1. An indication of the scale of a drama set in a small television studio. The acoustic effect within the set is likely to depend more on the scenery than the acoustics of the studio. Reflections from the studio walls tend to arrive too late to be of practical value.

2. Curves showing the reflective properties of various materials frequently used in the construction of scenery. Note that the reflectivity can vary considerably with frequency. This can cause colouration of the sound. (a) Thick fabric; (b) painted canvas flat; (c) wallpaper on plywood; (d) fibreglass on timber frame (shaped to look like brickwork).

3. Curvature of set may focus unwanted source of noise on to the live face of the microphone.

31

Noise can be a problem in video sound production.

Acoustic Isolation

A measure of the suitability of an area for sound production is the separation that is possible to achieve between the various sources of sound both wanted and unwanted.

Noise weighting

The definition of noise is 'unwanted sound'. Although it would be possible to measure the actual sound pressures involved, what is more important is to establish its nuisance value. To assess the effect it will have on the listener it is necessary to take into consideration the unequal sensitivity of our hearing and the way it changes with different sound levels. For this reason noise level meters are usually provided with three scales of 'weighting' corresponding to different phon contours (i.e. curves of equal sensitivity for different sound levels). Three switchable weighting networks are used according to the level of sound corresponding to the 40, 70 and 80 phon contours. The results are said to be A, B or C weighted.

In assessing the suitability of a room a series of *Noise Criterion* curves have been established. The generally-accepted standard for a small general-purpose studio is NC 15–20, rather less where the noise is easily identifiable or irritating in character. Small listening rooms should be better than NC 30.

Masking

Where noise is continuous it is important to take into account the masking effect whereby our hearing sensitivity is reduced for frequencies in the presence of other sounds of similar frequency. This is the basis of the Dolby noise reduction system (see pages 204–211).

Domestic situations

In normal domestic rooms noise tends to be air-borne (mainly through doors and windows) and conducted through the structure. Exterior noise can be reduced considerably by providing secondary double-glazing with the panes spaced 10–15 cm (4–6 in) apart and preferably not quite parallel to each other. Doors should be fitted with draught seals (ideally of the magnetic type as used in refrigerators) and cushioned thresholds. If possible double doors should be provided, forming a sound lobby, but otherwise the door could be covered by a thick drape. It may be necessary to carpet the room above and corridors outside the prevent the transmission of footfalls owing to sound conduction.

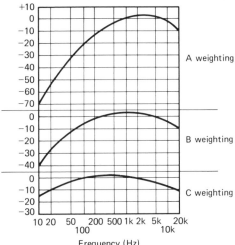

1

Frequency (Hz)

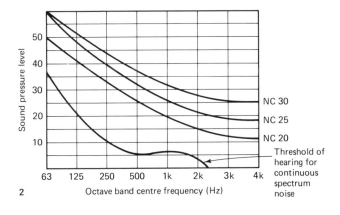

2

Octave band centre frequency (Hz)

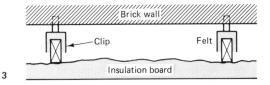

3

1. The response of weighting networks used in conjunction with meters for the measurement of noise. They are chosen according to the loudness level of the noise and take account of the variation in the sensitivity of our hearing at different intensity levels.

2. Noise criterion curves. Small studios should have a measured NC of 15–20. Listening rooms should be better than NC 30.

3. A simple way to improve insulation from noise through a wall by using insulation board fixed on battens and retained to the wall by felt-lined clips.

Microphones

Microphones convert acoustical energy into electrical power either by direct contact with a vibrating source or, more usually, by intercepting sound waves. Converting air waves into electrical signals involves the use of a diaphragm, a sensitive membrane which can respond to air pressure variations. There are two basic modes of operation: pressure and pressure gradient.

Pressure-operated microphones
In the pressure-operated microphone one side of the diaphragm is exposed to the air and the other sealed off, except for a small aperture which allows slow changes in atmospheric pressure to equalise on both sides of the diaphragm (like the eustachian tube in our ears). The diaphragm acts like a very quick-acting barometer to respond to minute changes in the surrounding air pressure. It is not concerned with the direction of the sound wave provided that its own dimensions are smaller than the wavelength of the sound. Pressure-operated microphones are *omnidirectional*, i.e. they pick up sound equally from all directions. Their output for a given direction of sound can be illustrated by a circle (ideally a sphere) drawn around a point which represents the microphone.

Pressure gradient microphones
Pressure gradient microphones have both sides of their diaphragms exposed to the sound waves, their output being produced by the difference in pressure (i.e. pressure gradient) between front and back at each moment in time, as a result of the different distance that the wave has to travel. As the path length difference between front and back of the diaphragm varies with the angle of incidence, the response of the microphone will also vary. In the extreme case, where the paths to front and back of the diaphragm are identical (as in most ribbon microphones) the directional response will be maximum at front and back, where the path length difference is maximum, and zero at right angles where it is equal. This is illustrated by a figure-of-eight polar diagram.

Combined operation
By combining the two modes of operation a variety of directivity patterns can be produced. If the figure-of-eight response of pure pressure gradient operation is combined with an equal output of omnidirectional response, one lobe of the figure-of-eight will be in phase with the omnidirectional output and will add to it, and the other lobe will be in antiphase and cancel it to zero. This produces the familiar cardioid (heart-shaped) directivity pattern. Increasing the proportion of pressure to pressure gradient operation creates a narrower cardioid front lobe with a small lobe at the back. This characteristic is known as *hypercardioid*.

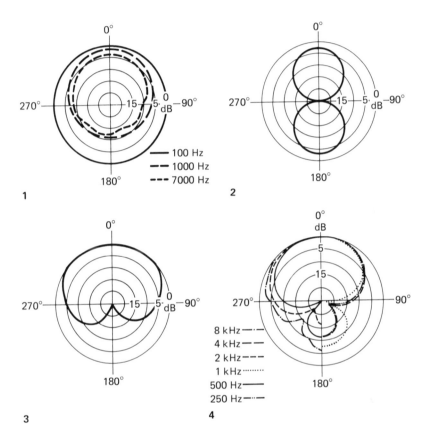

1

2

3

4

The directional response of microphones can be represented by a polar diagram. This is a circular graph in which the centre represents the microphone. The distance from the centre of a line drawn around it represents the sensitivity at each angle with respect to its front axis.

1. A perfectly omnidirectional microphone would have a circular polar diagram. In the one depicted this is only true for low frequencies (100 Hz). At higher frequencies the 'shadow' of the microphone case reduces the output.

2. The figure-of-eight characteristic is typical of a ribbon microphone.

3. Cardioid (heart-shaped) polar diagram. The cardioid shape can be obtained by adding together equal omnidirectional and figure-of-eight outputs.

4. Hypercardioid characteristic created by increasing proportion of pressure gradient element.

Choosing Microphones

There is no such thing as a 'universal microphone'. It is important to select one with the right characteristics for the situation.

Sensitivity

The output of the microphone must be sufficient to allow a large margin of signal above its inherent noise level and that of the following amplifier.

Microphone sensitivity is usually expressed as its open-circuit voltage at 1 kHz in mV/Pa (millivolt/pascal), where 1 Pa = 1 N/m^2 (newton per square metre). Note the threshold of hearing is approximately 2×10^{-5} N/m^2.

Typical sensitivities for dynamic microphones are 1–1.5 mV/Pa and for condenser mics (with head amplifiers) 5–10 mV/Pa.

Overload

Just as important as sensitivity for weak sources is the ability to withstand high sound pressure when used for close working or where the sound level is very loud (e.g. some pop group applications). Some condenser microphones are provided with attenuators which can be interposed between the capsule and head amplifier when they are to be subjected to high sound levels. Some have switchable attenuators built-in.

Physical characteristics

When microphones are to be seen in shot physical appearance and small size are obviously important. If they are to be hand-held they must be robust and not susceptible to handling noises and movement. Microphones for use in television studios and out of doors may have to withstand large ranges of temperature and humidity.

Frequency response

Most professional-grade microphones have frequency responses that vary only ±2–3 decibels over the full range of 20 Hz to 20 kHz both on and off axis. But this degree of fidelity is not always necessary, or even desirable, as sometimes an unequal response is particularly suited to a specific application.

Directivity

The choice of microphone directivity pattern depends on the acoustic situation and the need to discriminate between sources of sound. Experience with microphones shows that it is often more important to be able to reject sound from a particular direction (e.g. a PA loudspeaker) than to have the microphone pointing directly at the wanted source, so a clean cut-off is very desirable. Where it is required to discriminate against sound coming from all directions, such as reverberation, cardioid and figure-of-eight types have an equal discrimination factor of 3:1. The hypercardioid characteristic gives a factor of 4:1. Further improvement in

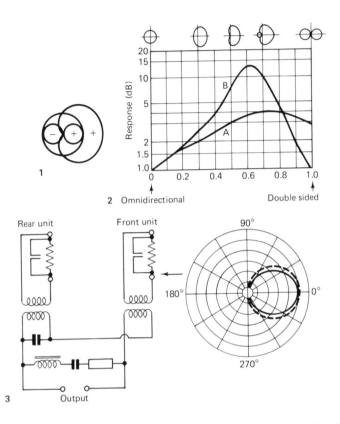

2 Omnidirectional · Double sided

1. The result of adding an equal amount of omnidirectional and figure-of-eight produces the familiar cardioid response.

2. Limacon curves show the effect of adding together different proportions of omnidirectional and figure-of-eight outputs. A, The result in terms of directivity (i.e. the ratio of incident to ambient sound). B, The unidirectional factor, i.e. the ratio of axial response to response from the rear. Important for isolation from PA and stage monitor loudspeakers.

3. The second-order gradient principle. Two ribbon elements spaced one behind the other with electrical correction in theory produce a single lobe. In practice this is very difficult to achieve over the whole frequency range.

incident/ambient discrimination can only be achieved by using focusing devices or by exploiting the second-order pressure gradient principle (see diagram). A narrow-angle response is often necessary in television to compensate for excessive microphone distances. Some video cameras are provided with 'zoom microphones' coupled to their zoom lenses, the idea being to match the sound to the picture, but there are limitations due to the variations in acoustic conditions and the possible disruptive effect of sudden changes in sound quality and background noise.

Moving coil microphones are robust, can have omnidirectional or unidirectional response and are of good quality.

Moving Coil Microphones

The moving-coil microphone consists of a diaphragm, which is usually a circular disc of aluminium alloy or plastic formed into a shallow dome with corrugations around the edge, close to where it is clamped, to allow it to move freely. To the back of the diaphragm is attached a coil of wire, usually wound from flat aluminium wire for lightness, which is suspended by the diaphragm in the annular gap between a powerful magnet and a central pole piece.

Induced voltage
When sound waves impinge on the front of the diaphragm, the back of which is partially or completely sealed off, the changing air pressure causes it to move in and out. This causes the coil, attached to the back of it, to move in the magnetic field, cutting the magnetic lines of force at right angles. This induces in the coil an electromotive force, which constitutes the output.

Microphone impedance
In order to reduce unwanted resonance of the diaphragm, the coil must be kept light and therefore consists of relatively few turns. Consequently the impedance of moving coil microphones tends to be low, characteristically about 30 Ω. Some incorporate a small step-up transformer into the stem of the microphone to increase the impedance to about 200 Ω. Some moving coil microphones have impedances that rise with frequency and need to be connected to amplifiers with an input impedance of about five times their own impedance to obtain a flat frequency response.

Omnidirectional dynamic microphones
Dynamic microphones that have the back of their diaphragms sealed from the air (except for a small pressure-equalising aperture) have an omnidirectional response for all frequencies with a wavelength longer than their dimensions. The larger types of microphone can create a 'sound shadow' which reduces the high frequency response off axis.

Directional dynamic microphones
Dynmamic microphones can be made directional by allowing a proportion of the incident sound wave to reach the back of the diaphragm through some form of acoustic delay. This can consist of a tube to increase the path length or some form of acoustic labyrinth such as a fine porous material. The acoustic delay produces a difference in the relative pressures between front and back of the diaphragm which varies with the angle of incidence. Thus, by careful design, a cardioid (single-sided) response can be obtained.

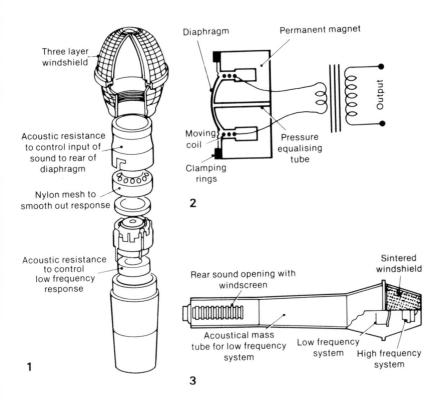

Three layer windshield

Acoustic resistance to control input of sound to rear of diaphragm

Nylon mesh to smooth out response

Acoustic resistance to control low frequency response

1

Diaphragm

Permanent magnet

Output

Moving coil

Pressure equalising tube

Clamping rings

2

Sintered windshield

Rear sound opening with windscreen

Acoustical mass tube for low frequency system

Low frequency system

High frequency system

3

1. Sectional view of a directional moving-coil microphone. Sound waves approaching from the front axis of the microphone reach the front of the diaphragm direct and the back via an acoustic delay so that a phase difference is created. Waves from the back encounter the acoustic resistance first, arrive at both sides of the diaphragm in the same phase and cancel.

2. Simplified diagram of the construction of a moving coil microphone. The diaphragm is dome shaped to promote stiffness.

3. A double element directional microphone. The problem of maintaining the proper phase relationship to achieve a cardioid response throughout the frequency spectrum is alleviated by dividing it into two with a separate transducer for each: one for the range 20–400 Hz and one for 400–18 000 Hz.

39

Pressure gradient microphones convert differences in pressure into electrical signals.

Ribbon Microphones

In the pressure gradient operated microphone both sides of the diaphragm are exposed to the sound waves. They make use of the difference in pressure that exists between the front and back of the diaphragm at any instant owing to the extra time that the sound takes to reach the back.

The most obvious example of the pressure gradient principle is the ribbon microphone. This consists essentially of a narrow ribbon of corrugated metal foil stretched between the pole pieces of a powerful magnet. As in the moving coil microphone (see page 38), movement of the conductor in the magnetic field induces electrical currents across the ribbon and these are applied via a step-up transformer (to increase the impedance) to provide the output. Sound waves have equal access to both front and back of the ribbon but waves approaching the microphone from the front obviously take longer to reach the back than the front of the ribbon. A difference of pressure thus exists at any instant in time owing to the difference in phase (time/wavelength relationship) between the two pressure waves (see page 14). This causes the ribbon to move away from the side of higher pressure.

Directional characteristics
Clearly the phase difference, and thereby the operating force, is greatest when the path-length difference is greatest, i.e. from front to back, and is non-existent when the path length is the same, i.e. sideways on. This gives rise to the familiar figure-of-eight polar characteristics of the ribbon microphone.

Unidirectional ribbon microphones
Some ribbon microphones are made asymmetrical between front and back of the ribbon by partially restricting access to the back or causing the sound to reach the back via a longer path. The effect is to reduce the sensitivity at the back of the microphone and, by adding to the forward pressure component, to increase the sensitivity in the forward direction. In this manner ribbon microphones can be given a unidirectional (cardioid or hypercardioid) response, the details and advantages of which are discussed on page 36.

Vertical axis response
Ribbon microphones have figure-of-eight polar characteristics in both the vertical and horizontal planes but they tend to be lacking in HF response off-axis in the vertical plane when the wavelength of the sound becomes comparable with the length of the ribbon, causing a wave motion along its length.

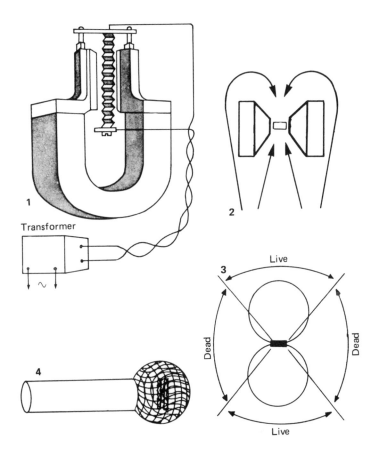

Transformer

1. Outline of a ribbon microphone showing the corrugated ribbon stretched between the pole-pices of a powereful magnet.

2. Plan view of ribbon microphone showing the extra distance that the sound has to travel to reach the back of the ribbon. The extra distance causes a time, and therefore a pressure, difference between front and back at any instant. This causes the ribbon to move.

3. The directional response of a ribbon microphone has a figure-of-eight polar diagram. Live front and back, dead at the side.

4. Cylindrical ribbon microphone. It employs a double ribbon at right angles to its length with an acoustic delay to the rear of the ribbon that give it a unidirectional response.

Condenser Microphones

The condenser microphone is really a small variable capacitor in which one plate is formed by the backplate and the other by the diaphragm, which is evenly spaced very close to it. Characteristically, the distance between the plates is about 25–40 micrometres (1–1.5 thousandths of an inch) giving a capacitance of about 10–20 pF.

Variations in sound pressure cause the diaphragm to move with respect to the backplate so that the capacitance varies in relation to the sound wave.

Polarising the capacitor

In order to obtain an output the variable capacitance must be turned into an emf. This is done by applying a polarising potential of about 50 V through a high resistance (100 MΩ or more). Variations in capacitance caused by movement of the diaphragm are thus converted into voltages across the resistor, which are then applied to a head amplifier (usually consisting of a field effect transistor) contained in the stem of the microphone. Some condenser microphones are provided with co-axial extension tubes which can be interposed between the capacitor capsule and the head amplifier.

Electret diaphragms

Some condenser microphones eliminate the need for a polarising voltage by having a permanent electrostatic charge, either in the diaphragm or the backplate (rather as a permanent magnet retains its magnetism). This results in a much more compact assembly, especially in the case of miniature clip-on microphones. Although a power supply is still required for the head amplifier this can be separated from the microphone capsule by a short lead.

RF capacitor microphones

Another type of capacitor microphone uses the variation in capacitance to beat with an RF oscillator, the rectified signal comprising the audio frequency output. This type of microphone can have an excellent frequency response and good sensitivity but can produce variable results when subjected to large variations in temperature and humidity.

Directional response

Capacitor microphones can be operated by pressure and are therefore omnidirectional if the back of the diaphragm is sealed (except for a small pressure equalising path). They can be made into very small capsules which do not produce sound shadows and thus retain their all-around response at all frequencies.

Alternatively the backplates can be made of a porous material which forms an acoustic delay path and produces a unidirectional response.

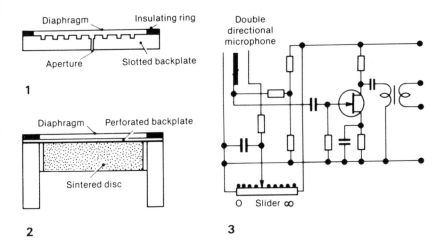

Diaphragm   Insulating ring

Aperture   Slotted backplate

1

Diaphragm   Perforated backplate

Sintered disc

2

Double directional microphone

O   Slider ∞

3

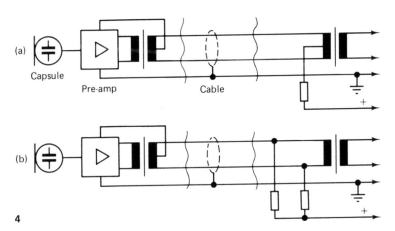

(a)

Capsule

Pre-amp   Cable

+

(b)

+

4

1. Sectional view of a pressure-operated capacitor microphone. The solid backplate is slotted to reduce the back pressure behind the diaphragm. The diaphragm is made of thin aluminium or metallised plastic stretched over an insulating ring. There is a small aperture to allow the pressure to equalise on a long-term basis.

2. Sectional representation of the essentials of a pressure-gradient (directional) capacitor microphone elements. The porous disc forms an acoustic delay.

3. A method of achieving variable directivity using a double-directional microphone, i.e. two diaphragms separated by a shared porous back plate. Moving the slider determines the relative polarity between the plates and produces a range of directivity patterns continuously variable between omnidirectional and figure-of-eight patterns.

4. Two types of phantom powering for condenser mics, to enable the head-amp to be powered via the microphone cable: (a) centre tapped transformer with common resistor, (b) artificial centre tap with two resistors.

43

Stand and Hand-Held Microphones

The use of a stand microphone in television suggests that:
1. The artist is static.
2. The microphone will come into shot.

A very important feature of a stand microphone, therefore, is its appearance. The microphone should be slim and neat and preferably mid tone in colour with a matt finish to minimise specular reflections from the lights. For the majority of circumstances a unidirectional response is required (either cardioid or hypercardioid) so that sufficient separation can be obtained from unwanted sound or accompaniment without the necessity to work so close that the microphone masks the artist or unduly restricts his movement. Experience suggests that 'end-fire' microphones are better for this purpose than side-facing ones because artists find it reassuring to have the microphone pointing toward them. (A side-facing microphone in a low position has to be angled away from the artist).

Microphone stands are normally telescopic and can be made or cut down to a variety of sizes. Normally, three sizes are useful for television: 0.3–0.5 m (1–1.5 ft) for low-speaking instruments such as the double bass and guitar amplifiers or for inconspicuous operation; 0.6–1.0 m (2–3.5 ft); for seated artists; 0.9–1.5 m (3–5 ft) for standing artists. These larger versions can also be used in association with small 'lazy arms' which are particularly useful for reaching across, for example, pianos or music stands. Extreme neatness and unobtrusiveness can be provided by the extension tubes available for electrostatic microphones (see page 42). These are thin co-axial tubes that can be interposed between the microphone capsule and the head amplifier, thereby presenting a very slim and unobtrusive appearance for the part of the microphone that is likely to come into shot. The tubes are curved at the top to provide an angle for the capsule which cannot be adjusted unless a special swivel joint is used between tube and capsule.

Hand-held microphones

When the appearance of a microphone in shot is not objectionable but the use of a stand would restrict the movement of the artist a hand-held microphone can be used.

The requirements for a hand-held microphone are that it should be slim, long enough to hold conveniently without the hand covering any sound entry ports, and as immune as possible from handling noise.

It should not suffer unduly from low-frequency rumble or wind noise when moved through the air and the cable must be sufficiently flexible so as not to cause rustles.

Some microphones, designed for hand-holding, have an additional transducer that produces an out-of-phase output to cancel handling noises (see page 58).

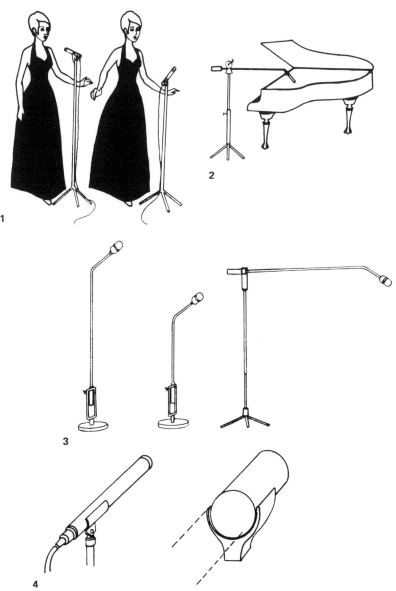

1. In general, end-fire microphones look better on stands than those with side sensitivity which, when used low enough to be unobtrusive, lean away from the artist.

2. Small lazy arm is especially useful for musical instruments.

3. Electrostatic microphones with extension tubes.

4. Microphone clip allows the microphone to be removed from the stand and hand held.

Slung Microphones

When it is required to cover a source of sound by means of a static microphone that does not appear in shot, it is generally best to place a slung microphone over the action. By the nature of things the source of sound, e.g. the artist's head, is usually closer to the top of the frame than any other edge.

The choice of a slung microphone suggests that the action is static. For fluid action several slung microphones are used, continually fading from one to another to cover movement or even changes in the angle of the artist's projection. This is not usually a satisfactory arrangement, because each microphone can produce a rigging problem and the risk of getting itself, or its shadow, in shot as the action progresses.

Microphones in shot
Slung microphones can look very ugly in shot and the fact that the point of suspension is above the frame of the picture can leave an uncomfortable query in the mind. They can be acceptable in long-shot providing the suspension is kept as neat as possible. Often the neatest method is to employ electrostatic microphones (see page 42) with the one-metre extension tubes hanging upside-down. If the lower end of these appear in shot at some distance it is likely to be very unobtrusive especially if it is covered with a dark stocking to avoid catching specular reflections from the lights.

An even neater arrangement is to hang an electrostatic personal mic upside-down by its cable, provided that directional qualities are not required.

Method of suspension
It is usually advisable to sling microphones from three points of suspension so that a wide range of adjustments can be made in the light of rehearsal experience.

Slings, or at least the last few metres at the microphone end, should be made of thin but strong material such as nylon or steel wire. Whenever possible elastic suspenders should be interposed between the sling and the microphone to prevent the transmission of sound along the sling reaching the microphone.

Stereo operation
When stereo is involved (see page 70) it is usual either to employ a special stereo microphone (two capsules in one case) or to suspend a metal bar about 15 cm (6 in) long on which two microphones can be mounted and set to the appropriate angle.

46

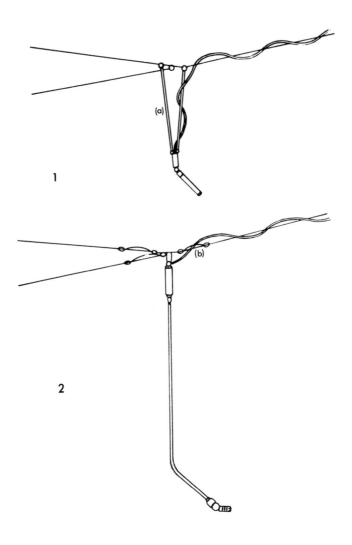

1. Whenever possible microphones should be slung from a three-point suspension so that they can be adjusted in all directions. It is always advisable to incorporate elastic suspension to prevent vibrations, transmitted down the slings, reaching the microphone. In the slinging method shown here, the rubber suspenders form the vertical elements and merely take the weight of the microphone.

2. An electrostatic microphone with extension tube slung upside-down makes a very neat suspension which is hardly visible in a long-shot. In the suspension shown, the suspenders are incorporated in the slings and have to take the full tensional strain. For safety reasons, therefore, they should be by-passed by slack nylon cords.

Hidden Microphones

If it is desired to cover some part of thet action by an out-of-shot fixed microphone but to sling it would result in its being too faraway, it is always worth considering the possibility of using a hidden microphone.

Maintaining microphone characteristics

The most important thing to remember about hiding microphones is that the material in which they are hidden must be acoustically transparent, i.e. sound must be able to pass right through it with a minimum of reflection or absorption. Otherwise, the response of the microphone is modified and the sound is so unnatural that the strategem is exposed.

Suitable material for surrounding a microphone is fine mesh metal or plastic gauze or fairly open-weave cloth. If this is light in colour and shaped to represent some solid article such as a vase or pile of books etc. the microphone can be hidden very effectively and its operation will be unimpaired. It is also possible to hide the microphone within a flower arrangement or in a lampshade etc. If there is a problem in hiding the cable, a radio microphone (see page 72) can be used, but care must be taken not to exhaust the battery by forgetting to turn it off between takes or rehearsals.

An ideal type of microphone for hiding would be the electrostatic personal type (see page 50), which can be very small and unobtrusive. They are omnidirectional so do not need to be angled carefully, but must be used at fairly close range to discriminate against noise.

The acoustical boundary type of microphone can be useful for hiding on tables and in some confined spaces such as vehicle interiors.

Sound perspective

Whatever type of microphone and hiding place is chosen, great care must be exercised to ensure that it comes roughly on a line between artist and camera and that the artist projects his voice in that direction to avoid reverse perspective effects (the artist appearing to recede in sound as he approaches in vision, and vice versa).

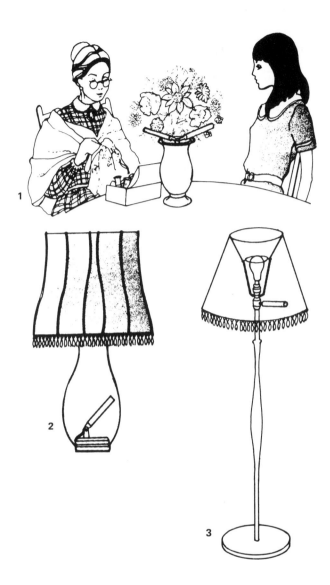

1. Vases of flowers can be useful for hiding microphones on a table, possibly between two people in dialogue. The two unidirectional microphones illustrated would be used so as to balance between the voices.

2. A table lamp can be constructed out of fine wire mesh covered with gauze. If it is light in colour it looks solid and the microphone is invisible.

3. Standard lamps can also provide useful hiding places, provided the shades are of acoustically transparent material.

Clip-On Microphones

One of the most convenient methods of picking up sound in vision is the clip-on or lavalier (necklace) microphone. Miniature moving coil microphones have been produced for the purpose, but the more expensive condenser types are much smaller and produce better results.

Moving coil clip-on microphones

Moving coil clip-on (lavalier) microphones have the advantage that they do not need a head amplifier and power supply, but it is essential that they are provided with very light and flexible cables to reduce rustling noises caused by movement of the clothing. It is usual to provide a junction to a heavier cable before it reaches the point where it is dragged along the floor.

Condenser clip-on microphones

Condenser microphones with electret diaphragms (see page 42) can be very small and inconspicuous. One current example is only 8.5 mm (0.3 in) long. They do, however, need head amplifiers and these usually take the form of a small cylinder which can be housed in a pocket or sling, connected to the microphone by a thin flexible cable. A heavier cable is then provided for the power supply and output of the amplifier. Alternatively the microphone can be connected directly to a miniature wireless transmitter (see page 72).

Using clip-on microphones

These microphones are so small as to be inconspicuous in shot and, being attached to the artist, overcome the problem of movement (especially when used with a wireless transmitter). The usual method of positioning the microphone, clipped to the lapel or tie, can produce quite good results provided that it is not shielded by the clothing — unless the material has an open weave such as a knitted sweater. But it is a very unnatural position from which to pick up the voice. There may be bass boom due to chest vibration, excessive nasal tones and lack of sibilants as the nose is in line with the microphone but the mouth may be shielded from the microphone by the chin. This can give rise to a distant 'off mike' sound which will vary considerably as the artist moves his head and will bear no perspective relationship to the picture.

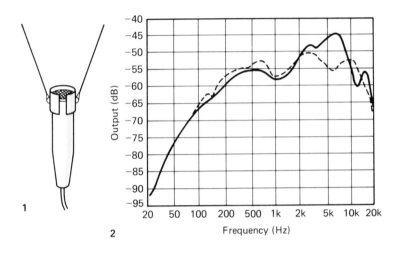

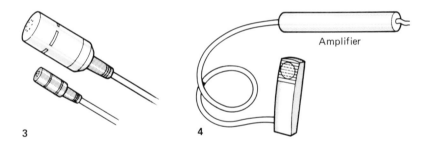

1. A personal lanyard microphone incorporating a movable clip.

2. Illustrating the effect of moving the clip. When the clip extends over the top of the microphone a cavity resonance effect causes an increase in the response at around 6 kHz giving increased 'presence'.

3. A miniature electrostatic personal microphone which uses an electret diaphragm, with a smaller and less conspicuous (through possibly less reliable) version.

4. An electrostatic microphone in which the diaphragm faces away from the artist wearing it. The idea is to give preference to the other person in an interview situation.

Super-Directional Microphones

There are many applications where microphones have to be used at a considerable distance from the sound source. These include ratio and television production covering sporting events and nature recording etc. The requirement is for a microphone with a very narrow acceptance angle to give maximum discrimination between the wanted source and the ambient noise and reverberation.

The reflector microphone

The simplest and most efficient method of producing a very narrow angle of acceptance is to place the microphone at the focus of a parabolic reflector. This arrangement is very efficient because the reflector acts like a large scoop which collects the sound and focuses a narrow beam on to the face of the microphone. The best type of microphone to use for this purpose is one with cardioid response arranged so that the live face looks into the reflector, encompassing all of it in its acceptance angle, and the back faces the wanted source of sound. This may seem odd until it is realised that it is a feature of the parabolic shape that the path lengths of all the waves from a source straight ahead to the point of focus are equal, so that they reach the microphone face in phase. Any sound picked up direct would travel a lesser distance and, being out of phase, would tend to cancel rather than augment the reflected sound.

There are two reasons for using as large a reflector as possible:
1. The larger the reflector the more of the sound wave it will collect and scoop into the microphone and thus the greater the forward sensitivity.
2. Surfaces can only reflect sounds of wavelength shorter than their dimensions (see page 14), so that for a parabolic reflector to operate through the full audio range it would appear to need to be of the order of 9 m (30 ft) in diameter!

In practice an acoustic effect comes to our aid. The sound pressure at the centre of a flat circular disc is greater than that in free air because sound waves are reflected at the surface and at the circumference of the disc. The reflected waves add at the centre of the disc and can increase the pressure by up to three times, i.e. up to 10 dB. As the microphone is near the surface most of this gain is available to maintain the directional response down to a much lower frequency than the dimensions would suggest. Practical parabolic reflectors are usually about 1 m (3 ft) in diameter.

Microphone in parabolic reflector. The cardioid microphone is mounted facing into the dish and away from the source of sound. Note sight hole and cursor for aiming the dish.

The angle of pickup of a parabolic reflector microphone depends on the sharpness of the focus, which also depends on the size of the diaphragm. With a small diameter diaphragm the angle can be so narrow that, unless the wanted sound source is at a considerable distance, it is necessary to defocus the microphone slightly. Parabolic reflectors are usually provided with gunsights to enable them to be aimed accurately.

When using parabolic microphones for sound effects at sporting events it is advisable to mount the dish in a high position, looking down, otherwise it is perfectly possible to pick up crowd conversation from the other side of the stadium.

Even a small parabolic dish can be extremely cumbersome, especially when used out of doors in a high wind.

The search for narrow-angle microphones to increase working distances has led to the development of the shotgun microphone.

The Shotgun Microphone

Unlike the parabolic reflector, which collects and concentrates the wanted sound, the shotgun microphone works by rejecting unwanted sounds and is no more sensitive than a conventional microphone.

Acoustic interference

The shotgun consists of a microphone coupled to the end of an acoustic interference tube. The tube has a large number of slots cut at right angles along its length. It works on the principle that sound arriving on the axis of the tube goes straight down to the microphone and is practically unaffected, whereas sound waves arriving from other angles enter through the slots and, according to the distance of each slot from the diaphragm, travel different distances and arrive at the diaphragm in differing phase, thereby tending to cancel each other out. Clever design of the acoustic damping in the tube and matching to the microphone helps to control standing waves and equalise pressures to promote maximum cancellation of ambient sound at the diaphragm.

Acceptance angle

As with all acoustic devices, interference tubes are only effective for sound with wavelengths shorter than their length. At lower frequencies the response reverts to that of the microphone to which the tube is attached, usually cardioid or hypercardioid. Most shotgun microphones have condenser (electrostatic) elements because of their high sensitivity and small size, but some are available which have high-output dynamic elements, especially suited to field work where robustness is a prime consideration.

It should be remembered that the interference principle is only really effective in free-field conditions and with plane waves (i.e. distant sound). It is not effective against reverberation, where the sound is coming from all directions in random phase. For this reason shotgun mikes work best in the open air or in television studios with very dead acoustics.

When shotgun microphones are used in windy conditions, or are moved about rapidly, they should be provided with windshields.

1

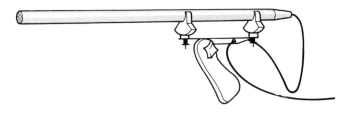

2

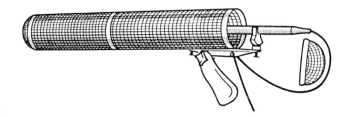

3

1. Shotgun microphone.

2. Shotgun microphone with pistol grip.

3. Shotgun microphone in windscreen.

Shotgun microphones are available in a variety of sizes from about 173 cm (5 ft 8 in) to 23 cm (9 in). The longer the better as far as directivity is concerned, but this has to be weighed against convenience, taking into account the purpose for which it is to be used. The longest shotgun tends to be cumbersome and its use is mainly restricted to distant effects coverage and audience participation programmes (see page 134).

Microphones with interference tubes about 40 cm (16 in) long are especially useful in association with vision. They can be hand-held with a pistol grip, or mounted on booms, fishpoles or cameras. They are effective at medium range, having an acceptance angle that varies between about 50° at high frequencies, widening as the frequency decreases until, at about 500 Hz, the tube ceases to be effective and the response reverts to that of the microphone element alone (cardioid or hypercardioid as the case may be).

Microphones with small interference tubes have less directional qualities but can be useful on fishpoles or hand-held for interviews, where they can provide reasonable separation from ambient sound without requiring accurate aim. They are also useful for isolation from PA.

Close-Working Microphones

It is a fundamental fact that microphones that work on the pressure gradient principle produce a marked increase in low frequency response when used close to a point source (e.g. a voice). There are two reasons for this:

1. At close range a point source produces a spherical wave front, which incrases the pressure gradient effect.

2. Following the inverse square law (page 16), the change in pressure over the distance between the front and back of the diaphragm increases at close range. These two factors combine to increase the pressure difference.

At normal working distances pressure gradient operation tends to favour high frequencies, because the phase change within a given distance increases with frequency. This is taken into consideration in the design of microphones with a flat response. The increase in response due to close working has much more effect at low frequencies, where the phase change is small and the pressure difference effect relatively greater. As a result the low frequency response increases progressively as the distance from the source is reduced.

Noise-cancelling microphones

The noise-cancelling principle is widely used for commentator's microphones in noisy situations and for musicians who sing and play their instruments (particularly drums) at the same time. As bass accentuation is a feature of pressure gradient operation the more the microphone relies on this element the more effective it will be. So the effect is most pronounced with figure-of-eight configuration, followed by hypercardioid. Some commentator's microphones have ribbon elements, with the ribbon shielded from the mouth by the magnet assembly, to guard against breath puffs and explosive consonants. The microphone is held at a distance of about 6 cm (2¼ in) from the mouth by a guard which is held above the upper lip. The fine wire mesh on top acts as a nasal windscreen.

Headgear microphones

Another type of noise-cancelling microphone consists of a miniature figure-of-eight or hypercardioid element mounted in a small boom arm attached to headgear, which holds it at the correct distance while leaving the hands free.

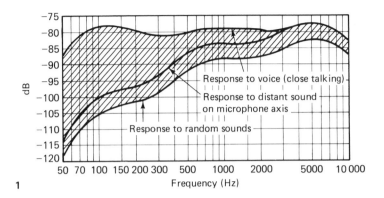

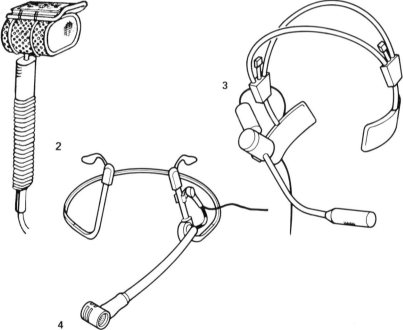

1. Typical frequency response curve (impedance = 30 Ω), illustrating the discrimination between close speech and distant random noise of a pressure gradient microphone. The discrimination is especially good for low frequencies, owing to the bass tip-up effect which effects close working. The equalisation provided in the microphone output levels the bass increase for the voice and in so doing makes the microphone less sensitive to background noise.

2. A noise-cancelling ribbon microphone. The bar at the top is held against the upper lip to fix the distance from the mouth. The handle houses a transformer.

3. A noise-cancelling microphone fitted to a headband.

4. A noise-cancelling microphone with a behind-the-neck band which is especially useful for vocalists who play instruments and move about.

Vocalist's Microphones

Microphones are available with characteristics especially suited for vocal work, but this is not to say that they may not have many other applications for which their design makes them appropriate.

Close-working vocalist's microphones

On the previous page we discussed the bass tip-up effect which occurs when a pressure gradient microphone is used close to a point source of sound. A vocalist's mouth can be considered to a point source of sound, located about 12 mm (½ in) behind the lips, so that the rise in low frequency response can be considerable when the microphone is only a short distance away, except in the case of some double-element microphones where the bass receptor gets its input from the rear of the microphone.

The close-up bass enhancement can be utilised in several ways. Vocalists can use the bass boost to extend their range for low notes, or the bass response can be equalised and the resulting loss for distant sounds used to improve discrimination against ambient sounds and feedback from loudspeakers etc. For this reason many vocalist's microphones are provided with a switch on their case giving such options as:
1. Full sensitivity, full frequency response.
2. Full sensitivity with cut-off below 100 Hz with about 12 dB/octave slope.
3. −14 dB sensitivity, full frequency response.
4. −14 dB sensitivity with roll-off of about 6 dB/octave below 500 Hz.

Some microphones also provide the option of 'presence boost' giving a peak of +2 or +4 dB at 4000 Hz, which accentuates the sibilants and tends to give the impression of closeness. These controls can be used to suit the microphone to the individual characteristics and power of the vocalist or instrument.

Vocalist's microphones normally have cardioid or hypercardioid characteristics, the latter being especially suitable for guarding against feedback from monitor loudspeakers (sited on the floor to help the artist to hear himself) and PA.

Hand-held microphones

Microphones for hand-holding must be robust. Some are shock-mounted within their cases and incorporate double breath screens. One type employs two transducers. One picks up the sound and the other is sensitive only to impact and shock noises, which it combines with the other in antiphase to cancel them out.

1. Graph illustrating the increase in bass response of a microphone as the distance from the mouth decreases. Even with the microphone 15 cm (6 in) from the mouth (which is greater than the distance usually employed by pop vocalists) the bass is increased by about 16 dB at 50 Hz.

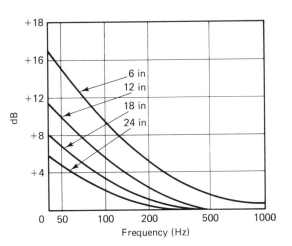

2. Graph based on the inverse square law, showing how rapidly the output from a microphone can rise (P2) with a comparatively small decrease in distance (A–B) when very near the source. This factor and the one above illustrate the vast range in volume and change in quality that can occur with very close working. Compare the small increase (P1) resulting from the same movement (C–D) at a greater distance.

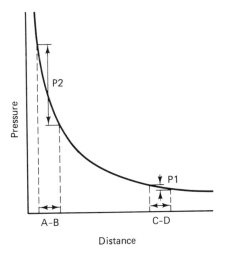

3. Section of microphone casing showing bass control and presence boost switching.

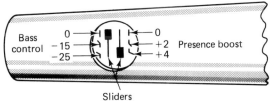

Acoustical Boundary Microphones

The acoustical boundary microphone consists of a small microphone element (usually self-polarised electrostatic) mounted close to, or recessed into, a baffle which can be placed on a table or floor or fixed onto any large reflecting surface. The principle of operation exploits the pressure-doubling effect that occurs close to a large sound-reflecting surface, owing to the sound waves effectively doubling back on themselves. If we imagine a sound wave encountering a rigid surface which reflects it back along its original path, it is evident that there can be no movement of the air particles past that point. This is effectively a node (zero point) as far as the particle displacement is concerned. But the pressure wave is maximum when the particle displacement is zero, so there are pressure wave anti-nodes (maxima) at the boundary surface. Moreover the reflected waves are returned in phase with the incident wave so that the resultant pressure will be doubled.

All this assumes that (a) the surface is perfectly reflecting, (b) it is rigid (i.e. the incident sound wave does not cause it to vibrate) and (c) the reflective surface is larger than the wavelength of the sound. If these conditions are not completely met the reflected wave will be smaller and possibly slightly out of phase with the incident wave so that the combined effect will be correspondingly smaller.

Boundary microphones are produced which consist of pressure elements recessed just below the surface of a small metal or wood baffle. These have an hemispherical response, i.e. half omnidirectional, with an output sensitivity 6 dB higher than a conventional omnidirectional microphone in the same position. An alternative version employs a small microphone capsule raised just above the surface of the capsule with the choice of half-spherical or half-cardioid response. The latter can be helpful in reducing the risk of PA feedback in some circumstances, such as when used on stage. Some have built-in filters to suppress footfalls and other low-frequency vibrations.

One application could be to mount an acoustical boundary microphone in a baffle of about 0.6–1 m (2–3 ft) diameter and suspend it vertically in front of a large sound source, such as a chorus. This arrangement would provide wide-angle coverage with a good degree of separation from PA etc. The baffle could be made of transparent material, e.g. perspex, to reduce the visual obstruction.

1

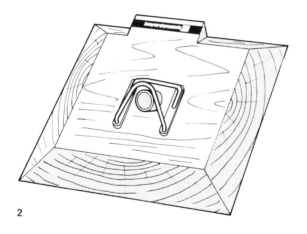

2

1. One type of acoustical boundary microphone consists of a metal disc 160 mm in diameter and 6 mm thick (6.3 × 0.25 in). A condenser element is completely recessed into the centre of the disc, giving a half-spherical response.

2. Another type has the microphone mounted in a slanting position just above the surface of a 220 mm (8.5 in) square wooden plate, protected from damage by a metal bow. A choice of capsules is provided to give half-spherical or half-cardioid response.

61

Windscreens

Most microphones are affected by wind. The more directional they are, i.e. the more they rely upon the pressure gradient principle, the more susceptible they are to wind.

It is an unfortunate fact that wind on a microphone does not make the sort of sound normally associated with wind, which might at least explain its presence, but a thumping or low rumbling sound — more like thunder.

There are two major causes of wind noise: (a) rapidly moving air actually hitting the diaphragm and (b) turbulence around sharp contours of the case, causing eddies to occur which penetrate the shielding to reach the diaphragm.

The diaphragm can be protected from the effect of wind blast by screening it with an acoustically porous material such as polyester foam or several layers of fine mesh. Such a windscreen has much less effect on the sound than it does on the wind, because the unilateral velocity of even a light wind of 10 km/h, 2.8 m/s (6 miles/h, 9 ft/s) is much higher than the vibratory velocity of sound waves (less than about 8 mm/s or 0.3 in/s) and the flow resistance of porous material increases with particle velocity. Nevertheless there will be some degradation in the sound quality (usually a reduction in HF sensitivity) when windshields are in use.

The effect of turbulence can be greatly reduced by creating a smooth airflow over the surface, which involves increasing the radius of the curves ideally to a spherical or cylindrical/spherical section. Unfortunately, to be effective against strong winds windscreens must be large in relation to the microphone; that can mean diameters of 10–15 cm (4–6 in), which tends to preclude their use in shot.

Pop windscreens

When speakers work close to the microphone there is a risk of breath puffs, particularly from 'explosive' consonants (P and B) which can produce air particle velocities of over 90 m/s (300 in/s) resulting in very loud thumps in the sound. Some microphones have close-talking shields built in to their construction, usually layers of fine mesh or 'sintered' material. Otherwise small windshields are available made of plastic foam supplied in various colours to match the costumes.

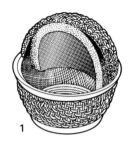

1

3

2

1. Section through an integral windscreen for a vocalist's microphone, which might be subjected to rough usage as well as close working. Two layers of strong wire mesh and one of fine wire mesh are followed by a layer of plastic foam. The microphone element is also resiliently mounted to absorb shocks and handling noise.

2. A selection of plastic foam windscreens to suit different types of microphones. They are available in a variety of colours.

3. Windscreens for gun microphones, showing end cover removed and additional 'windjammer' cover, which can be zipped over the outside to protect against severe wind.

There can be no hard and fast rules about the choice and positioning of microphones, but here are some general points.

Using Microphones

How many microphones?
In general the fewer the number of microphones in use the better because, unless each source is confined to its own microphone, the overlapping sound fields will reduce the clarity of the sound. Nevertheless there are many circumstances where the desired quality of sound can only be achieved by multi-microphone technique. At one end of the scale it is possible, given the right conditions, to record a large classical orchestra on a pair of microphones, while in rock music it is quite common to use 10 microphones on a drum kit. The factors that govern the number of microphones to use are the internal balance of the source, the closeness of the technique required to produce the desired quality of sound, and whether different elements in the balance require different degrees of processing (frequency response shaping, reverberation, 'sweetening' etc.).

Internal balance
Very few programme sources are sufficiently well balanced within themselves to meet the limitations of reproduction in the domestic situation (see page 20). Even in the case of the classical orchestra, mentioned above, the conductor would expect to receive guidance on points of musical balance from the sound balancer, because from the conductor's position it is not possible to judge the instrumental balance with sufficient accuracy for recording purposes. The majority of musical combinations play orchestrations written with little attempt to achieve internal balance, on the assumption that each section or individual musician will have a separate microphone which can be controlled to obtain a balanced mixture at each moment in time. Similarly, in drama or actuality situations a great deal of selection and artistic control is required to obtain an effective production.

Sound character
The character of the sound depends upon the type of microphone, its proximity to the source and the technical processing applied to its output. The main factors governing closeness are the ratio of incident to ambient sound, i.e. separation from reverberation and other sources, and sound quality. Most instruments produce differing tonal characteristics from different directions and degrees of closeness. To obtain the sort of crisp attack required, particularly in association with close-up pictures, a close technique is necessary. Reverberation can be added artificially.

1. A hypercardioid microphone to give maximum discrimination against PA loudspeakers.

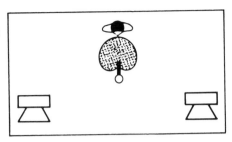

2. A cardioid microphone for a vocalist angled so as to discriminate against the orchestral accompaniment. The angle of discrimination of a cardioid microphone is much more acute at the back than at the front. The small 'fold-back' loudspeaker may be necessary to enable the vocalist to hear himself.

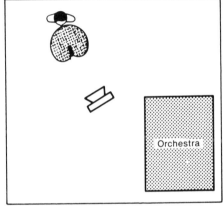

3. The figure-of-eight characteristics of a ribbon microphone used to isolate the comparatively weak sound of a flute against the powerful tones of a trumpet.

In selecting the type and position of a microphone for a particular purpose it should be remembered that it is often not so important to point the live side of the microphone directly towards the wanted source as the dead side towards the source to be discriminated against.

When using more than one microphone it is important that they are properly phased. This can be checked by placing them near to each other and speaking between them. If they are out of phase the sound will become distorted and weaker when they are both faded up.

Stereophonic Sound for TV

Stereophonic sound is becoming increasingly available, particularly on video cassettes. Stereo-capable television sets are available and broadcast transmission systems are being brought into service, particularly via satellites.

Binaural hearing
Most people are able to listen with two ears, which gives us the ability to judge the direction of a source of sound and, to a certain extent, to discriminate between sounds coming from different directions. When sound comes from a direction other than straight ahead it reaches the nearer ear slightly before the other one. The brain is able to detect this tiny difference in time and also the slight difference in intensity and quality (due to the waves being deflected around the head) and interpret it as information about the direction of the sound source. We can also tell whether the sound is coming from in front or behind, above or below, by slight differences in quality resulting from the shape of our ear lobe (pinna), which has a slightly different effect on sounds from different directions.

Aural discrimination
The ability to locate the direction of sound sources can enhance the realism and clarity of the reproduction, while the ability to distinguish between direct and reverberant sound can give a sense of depth and third dimension.

Stereophony
Stereophony is a two-channel sound system with lateral information which gives the effect of a 'sound stage' between two loudspeakers. With normal audio recording techniques it is most effective when the two loudspeakers are about two metres (six feet) apart and an equal distance from the listener.

Stereophony for television
It would at first appear that stereo sound is not appropriate for television as effective stereophony would produce a 'sound stage' that is much larger than the television screen. Also the vision direction (influenced by the small screen) tends to keep the centre of interest in the centre of the screen and to illustrate distance and separation in depth rather than in width.

Nevertheless there is no question that stereo sound is superior to mono in terms of realism and impressiveness and in its ability to convey the illusion of a third dimension.

1. For effective stereophony the listener should be equidistant between the loudspeakers and at an angle of about 60° to each.

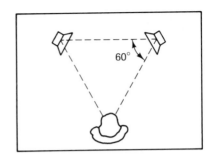

2. Typical stereo TV receiver. The narrow spacing of the loudspeakers gives limited stereo effect, but still provides enhanced depth and some degree of separation.

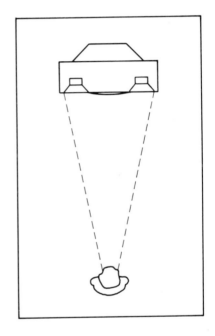

3. Improved arrangement for stereo reception. Increased width improves stereo imaging and time delays resulting from reflections from walls enhance the effect of ambience.

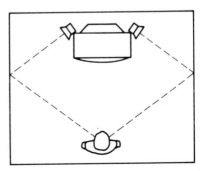

Applying Stereo Sound to Vision

It has to be admitted that few people will listen to television sound with a pair of well-matched loudspeakers placed about one metre (three feet) on either side of the television set and equidistant from the viewer. Stereo-equipped television sets mainly have loudspeakers closely spaced at each side of the screen. So in applying stereo sound to television it is important not to dwell too much on the apparent position of the sound but to concentrate more on those aspects that will be effective under less than ideal conditions and are not too dependent on the size of the screen — i.e. the sense of depth and spaciousness, perspective and separation between solo and accompaniment, effects and atmosphere etc.

When watching television we are aware that most of the time we are only seeing a small segment of the sound source. Just as in real life, when we concentrate our gaze in one direction we cannot shut our ears to the sounds that are going on around us, so it is not unnatural for the sound image to extend beyond the confines of the television screen.

Stereo sound can be most effective in increasing the distinction between sources of sound seen in shot and offstage sounds and effects. These can be placed to one side or the other as appropriate or, like atmosphere sounds, musical accompaniment and reverberation, spread across the sound stage. However, for broadcast purposes it is important not to place the offstage sounds so far off centre that they lose impact for the listener in mono (at present the vast majority).

One way of applying stereo sound to television is to consider it in terms of 'incidence and ambience'. In other words, arrange for sound sources that appear on the screen to have a reasonably central image (within about 15°), and fix everything else that does not have an obvious geographical relationship to the picture to be just 'elsewhere', i.e. indeterminate in position.

Surround sound

The effect of ambience can be greatly enhanced by applying 'surround sound', using several loudspeakers, suitably phased and spaced around the room to reproduce the 'off-stage' sound while the picture-related sound comes from the direction of the screen.

The positioning of 'ambience' loudspeakers can be made less critical if they are fed via a delay of about 20 ms. This ensures that the sound does not reach the listener before the sound from the TV set, thereby preventing on-screen sound from appearing to come from the surround-sound loudspeakers. (See Haas effect, page 116.)

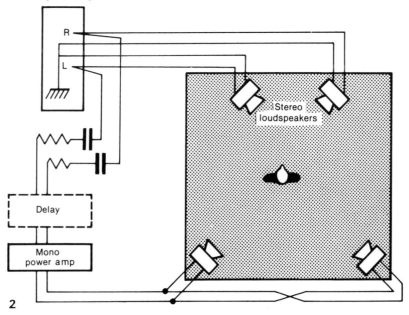

1

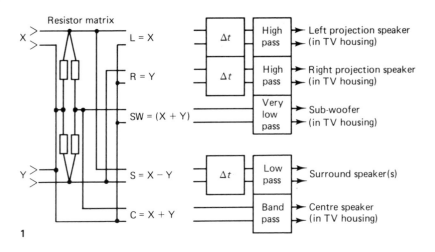

2

1. Basic Dolby surround sound decoding and typical processing arrangement. A new fast real-time processor can simulate sounds with variable distance and position.

2. A simple method of obtaining a form of surround sound. Most stereo power amplifiers have one leg of each loudspeaker connection common to earth. Connecting between the two live terminals therefore produces an antiphase output (R−L). This can be applied to a mono amplifier connected to two anti-phased loudspeakers.

Specialised microphone technique is required for stereo sound.

Microphone Technique for Stereophony

Microphone technique for stereo usually involves a laterally spaced or coincident pair of microphones to establish an overall sound image, to which can be added further stereo pairs or single microphones, the outputs of which can be pan potted into their respective positions in the 'sound stage'.

Pan pot

The pan pot (panoramic potentiometer) is a pair of ganged potentiometers with which a sound source can be differentially divided between the stereo channels to place it in the desired location in the sound image.

Stereo microphones

The two microphones forming a coincident pair can be mounted on a bracket facing each other at right angles with the capsules close together. Alternatively a microphone with two capsules in a single casing can be used. This type is usually a double condenser microphone with one capsule capable of being varied in angle with respect to the other. Various combinations can be used, such as two cardioids, or figures-of-eight, or a cardioid and a figure-of-eight, or a figure-of-eight and an omnidirectional.

The X–Y system

If the two mics on the bracket or in the coincident pair are set at right angles (45° to the centre of the sound source) and fed to their respective left (A) and right (B) channels, this is known as the X–Y system. In stereo it provides a 'sound stage' image spread between the two loudspeakers. The simple addition of the two signals provides a compatible mono signal, provided that they are kept in perfect phase throughout the system.

The M–S system

An alternative arrangement is the M–S technique. An M–S pair could consists of a cardioid microphone facing forwards, with a figure-of-eight at right angles. The forward-facing element provides the M (middle) signal or centre image and the other the S (side) signal. The M–S arrangement is very suitable for the boom as it provides a very clean well-defined stereo image, the width of which can be adjusted by altering the relative proportions of M and S.

The left and right (A & B) signals can be matrixed from M & S; the sum of M & S makes A and the difference makes B. Similarly M and S signals can be matrixed from A & B by $A + B = M$ and $A - B = S$.

Adding the two signals for centrally placed coherent sound will produce double the output (+6 dB). If the sound is not co-related the combination will give an increase of 3 dB more than each microphone on its own. In practice, particularly where coincident pairs are used, there is consider-

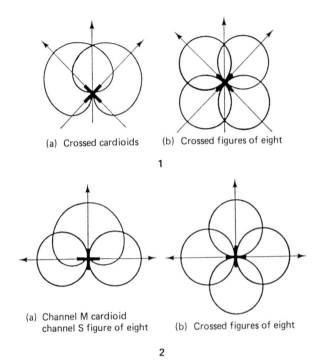

(a) Crossed cardioids   (b) Crossed figures of eight

1

(a) Channel M cardioid
    channel S figure of eight   (b) Crossed figures of eight

2

Different response configurations for coincident microphones.

1. Two methods of arranging coincident mics in the X–Y system.

2. Alternative methods of producing the M–S signal.

able co-relation between the signals, but for compatibility with the derived mono it is common to line-up the mono signal 3 dB below the stereo combination.

For television a typical approach would be to employ a coincident pair in the boom, augmented by the output of additional 'spot' mics pan potted into their respective positions. If an orchestra or other large source of sound is involved, other stereo pairs could be added. Ideally the additional mics should have delays in their outputs to prevent phasing problems which could blur the stereo image.

Radio Microphones

When an artist has a large area of movement, or when a cable would be an encumbrance, a radio microphone can be used. This usually consists of a small transmitter, carried on the person of the artist, connected to a personal microphone (either clip-on or hand-held). The transmitter is usually carried in the pocket or in a pouch tied around the waist. There are also radio microphones in which microphone and transmitter are built into the same case in the form of a baton to be held in the hand.

Transmission systems

The frequency-modulated signal is picked up by a receiver which is normally housed in the sound control room. The choice of receiving aerial depends on the circumstances. If only one artist, with a limited range of movement, is involved, a single dipole aerial can be used, mounted as near as practicable to the action. If several artists with radio microphones are required to move about extensively, it may be advisable to work their respective aerials at some distance (e.g. 10 metres or 30 feet away from the action) and well separated from each other to avoid 'frequency pulling' of the receivers by the other transmitters.

Long range reception (e.g. an operator walking with a shotgun microphone and transmitter to cover sound effects for a golf match) may require a highly directional aerial panned to follow the action. In difficult circumstances, where large areas of movement are involved and there are multiple reflections from moving cameras and other metal structures, a form of 'diversity reception' is required to avoid the effect of 'nodes' (positions of signal cancellation) due to out-of-phase reflections. This can be a simple matter of combining a number of aerials into one receiver, or a system of switching between two aerials at a supersonic rate, the theory being that the signal should be available from one or other of the aerials at any given instant.

The receiver should have a limited AGC or a muting circuit to prevent sudden increases in noise level in the absence of signal, which is much worse than momentary silence. In the ultimate a sophisticated diversity receiver would be used which measures the output from several aerials and automatically switches to the best one from instant to instant or combines their outputs. When a number of radio microphones are to be used in close proximity they can tend to pick up each other's signals and re-radiate them as intermodulation products. So it is important to choose frequencies which do not coincide with the sum or difference frequencies of any of the others (i.e. $2f_1 - f_2$ or $2f_2 - f_1$).

Site surveys

Watch out for large transmitters in the vicinity, police and other radio intercom systems, neon lighting and other sources of electrical interference, such as HF noises due to digital appliances (synthesizers and effects units etc.). Filter units are available which should be connected in the inputs and outputs of such equipment if they cause interference.

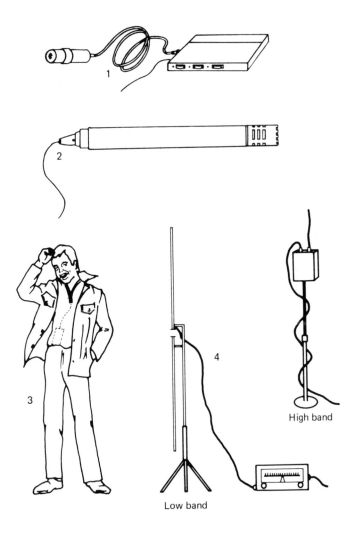

1   2   3   4

High band

Low band

1. Radio microphone for studio use. The transmitter is miniaturised and made as inconspicuous as possible. The microphone can be any of the personal or hand-held types. The miniature clip-on type is especially suitable.

2. Baton microphone for interviews. This type can be handed from person to person easily.

3. Method of accommodating a radio microphone.

4. The aerial should be as close as possible to the action. A portable stand can be useful. For theatre work it is often possible to get the aerial close to the artist by positioning it under the stage.

Fishpoles

In circumstances where the action covers a limited area and microphones are not to be seen in shot, but the use of suspended or hidden microphones would be awkward or too inflexible, 'fishpoles' can be useful. These can consist of light aluminium or carbon and glassfibre poles with telescopic sections allowing for extension between about 1.25 m and 4 m (4–13 ft), with a lightweight microphone mounted on one end.

Type of microphone

The best type of microphone for this purpose is usually an electrostatic cardioid, which can combine light weight with good directional characteristics. However, if the fishpole microphone is used only for a short sequence to take over from some other method, such as a boom, it might be necessary to employ the same microphone type as used on the boom to preserve continuity. Electrostatic microphones are often used on booms so there should be no problem in matching the types, which can include short shotgun microphones.

Microphone suspension

The microphone should be suspended in a suitable mounting and if necessary provided with some windshielding. The degree of windshielding required depends on whether the action is out of doors with a possibility of real wind or if it is necessary merely to guard against wind turbulence due to the microphone being moved rapidly through the air. In this latter case a relatively small close-talking shield is adequate. Protection against strong wind requires a much larger windshield, possibly in excess of 10 cm (4 in) in diameter, to reduce the effect of minor turbulence over its surface reaching the diaphragm. It is important that the microphone is rigidly clasped inside the windshield and the shock suspension applied overall, otherwise movement inside the shield as the pole is swung about would cause its own air turbulence.

Operation

Operator fatigue can be reduced by providing a counterweight at the other end of the pole to form a balance at the point of holding. If the sequences are longer than a few minutes and the rod has to be held above the action a shoulder bracket can be an asset. One of the main advantages of the fishpole, particularly when used in confined spaces, is the ability to use it below or to the side of the shot.

Shotgun mics, with pistol grip suspensions, can be excellent for fishpole operation especially when the optimum approach is from the side or under the shot. The use of a radio transmitter with a fishpole or hand-held line microphone can greatly increase operator mobility.

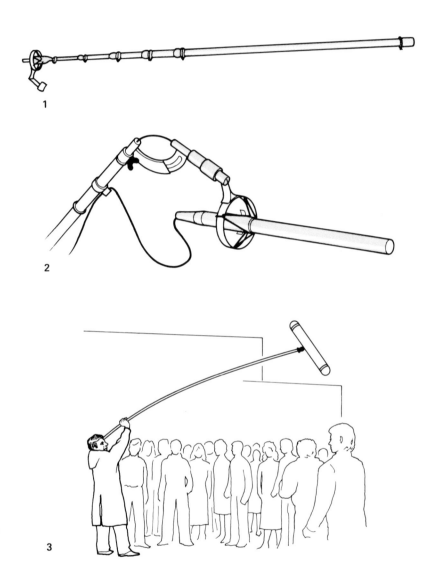

1. A telescopic fishpole, extendible between 1.25 and 4.2 m (4 ft 1 in–13 ft 9 in).

2. A shock-absorbing mounting for use on a fishpole.

3. A shotgun microphone in use on an extended fishpole. It is held above the head with arms extended to clear shots.

Small Booms

The fishpole, described on the previous page, can be a very convenient method of suspending a microphone close to the action without it coming into shot. However, it does have two serious limitations:

1. Limited reach, which can inhibit its use where long-shots are interspersed with close-ups.
2. Operator fatigue, since it is usually necessary to position the microphone above the picture, which requires the operator to hold the pole at arms length above his head. This position can only be maintained for quite short takes.

The problem can be overcome by the use of a small boom in the studio situation if the action is reasonably static. A simple boom consists of a tripod on wheels, with a central column which can be adjusted for height. On top of this is pivoted the boom arm, which consists of a telescopic pole which can be extended to about 4 metres (13 ft). The back end has a counterbalancing weight which can be slid along the pole to compensate for varying weights of microphone and length of arm.

Operation

The boom is used to suspend the microphone in front of the action and just above camera shot. It can be raised or lowered to accommodate different widths of shot by manipulating the back end. In the same way it can be swung from side to side to favour different artists or accommodate their movement about the set.

What is much more difficult with this type of boom is to compensate for backward and forward movement of the artists. The tripod base is on wheels and it is fairly easy to track the whole thing forwards while watching the action and maintaining the correct microphone height, but it is much more difficult to track backwards without running over the cable or into other obstructions.

It is at all times necessary to study the position of the lights to ensure that the microphone or boom arm does not cast shadows on the artists or that part of the background that is in shot.

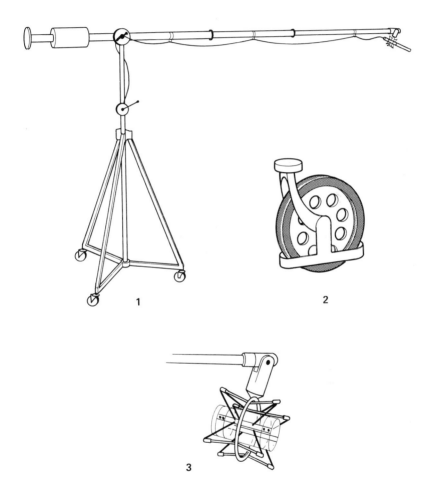

1. Small boom.

2. Detail of improved castor, with large wheel for easy tracking and cable guard.

3. Detail of suitable shock-mounting to prevent transmission of rumble along the boom arm, particularly when it is tracked. Also to soften microphone shake.

The microphone boom is the basic tool of television sound.

Sound Booms

There are several types of microphone booms but essentially they consist of a long telescopic pole with a microphone on one end and a counterweight on the other. This is pivoted on a vertical column on 'the pram', which is mounted on wheels and usually provides a raised platform and in some cases a seat for the operator to see the action above the normal height of the cameras.

Operation

The length of the arm can be adjusted by means of a 'racking handle' which operates through a series of pulleys. The weight of the microphone on one end of the boom arm is counterbalanced by a much heavier lead weight at a lesser distance from the fulcrum on the other end. As the telescopic arm is racked out the counterweight moves back a corresponding amount so that the balance is uniformly maintained. In addition to racking the microphone in and out, the operator can also swing the arm through an arc and swivel the microphone by remote control in both the horizontal and vertical plane. This is achieved by a handle, held in the left hand, which is raised or lowered to twist the microphone sideways and squeezed to tilt it up and down. The handle is mounted to the rear of the pivot point, while the handle operating the rack mechanism, which is operated by the right hand, is in line with the pivot so that all four actions can be accomplished simultaneously.

The boom pram

The boom pram is normally fitted with three wheels; some are capable of being crabbed (i.e. all wheels steered parallel to each other) or tracked (the rear wheels only being steered). Some types of boom provide the ability to alter the height of the arm fulcrum and the platform (the relationship between these two must be fixed) either by a winding mechanism or by hydraulic operation. This latter arrangement enables the adjustment to take place even during the action, making it possible to dip down to track under a low door or to rise above a camera shot.

Typical boom dimensions are as follows:

Pram width 1.2 m (4 ft), length 1.5 m (5 ft). Boom arm maximum extension 6 m (20 ft). Tail length 1.2 m (4 ft). Minimum extension 3 m (10 ft). Maximum microphone height 4.5 m (15 ft). Minimum overall height 2.5 m (8 ft).

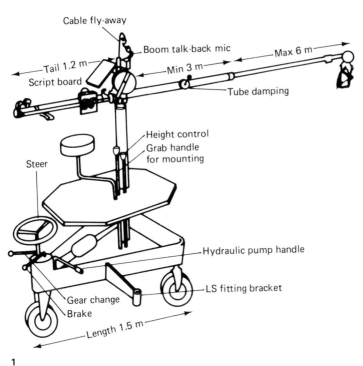

Cable fly-away

Boom talk-back mic

Max 6 m

Tail 1.2 m

Min 3 m

Script board

Tube damping

Height control
Grab handle
for mounting

Steer

Hydraulic pump handle

LS fitting bracket

Gear change
Brake

Length 1.5 m

1

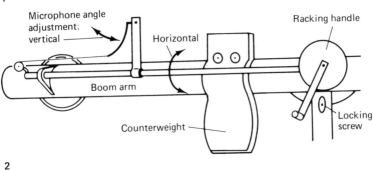

Microphone angle
adjustment:
vertical

Racking handle

Horizontal

Boom arm

Locking
screw

Counterweight

2

1. A boom suitable for most studio applications. The height of the platform and arm is adjusted by an hydraulic ram controlled by a lever on the platform. The revolving seat is intended for occasional use. It is difficult to do rapid swings and complicated manoeuvres without standing up.

2. Detail of back end of boom arm showing operational control.

Boom Microphones

The most important point to remember when selecting a microphone for use on a boom is that for most of the time the microphone will be operating at a distance greater than the optimum.

Directional characteristics

The directional characteristics of the microphone are of special importance as there is a continual requirement to discriminate between wanted and unwanted sound such as noise, orchestral accompaniment (which should be picked up only by microphones properly positioned for the purpose), and reverberation etc. This discrimination between wanted and unwanted sound must be good enough to allow the microphone to work at a sufficient distance so as not to cramp the camera headroom.

Moreover, as it is inevitable that a proportion of the accompaniment will be picked up on the dead side of the microphone, it is important that it has an even off-axis response, i.e. that the 'dead' areas are equally 'dead' over the full frequency spectrum. This is especially true in the case of music where, unless the microphone can be used sufficiently close to the artist to achieve discrimination by exploiting the relative distance of the sources, a proportion of the accompaniment will be picked up on the dead side of the microphone. In these circumstances it is usually worthwhile to employ a good quality electrostatic microphone for the boom provided that a suitable shock mounting and sufficient wind shielding can be provided.

Acceptance angle

The most useful acceptance angle for a boom microphone is likely to be cardioid. This gives a reasonable compromise between the narrow response necessary to give sufficient discrimination at a reasonable working distance from the artist and the ability to pick up dialogue from artists spaced apart by several metres without making the aim too critical.

Although the microphone can be tilted and turned in all directions there is seldom time between speeches to re-aim it accurately. Where exceptional working distances are involved, or the need for separation is great, e.g. in audience participation programmes (see page 134), shotgun microphones (see page 54) can be used.

Construction and mounting

Microphones used on booms must be robust to stand up to continual shaking. Even so they must be provided with effective shock mountings. Effective windshields are also necessary if boom microphones are to be moved rapidly, and especially if they are to be used out of doors.

Stereophony

Stereophonic microphones, of the type in which both capsules are contained in the same casing, are ideal for boom use.

The most useful configuration is the M–S format (page 70).

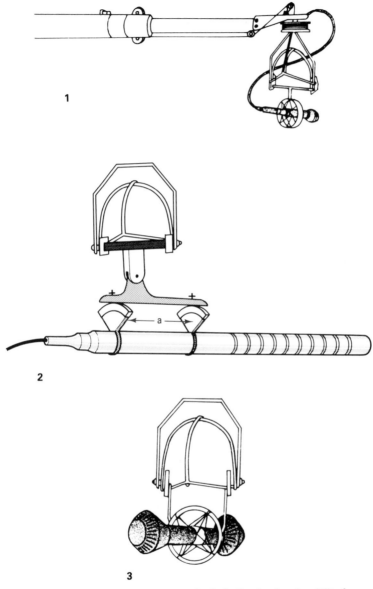

1. Detail of end of boom arm showing method of adjusting 'pan' and tilt of microphone. The microphone is mounted in rubber suspenders within the tilting cage which is also mounted on soft suspension.

2. Shotgun microphone in boom mounting with soft rubber suspenders (a). If the boom is to be swung rapidly, or used in the open air, a windscreen must be fitted.

3. Double element microphone in use in boom showing the need to fit windshields at both ends.

The Scope of the Boom

In every production situation there is almost certainly one optimum position for each microphone. If a microphone is to be used to follow free-flowing movement, it too should be mobile and the best way to achieve this is with a microphone boom.

Advantages

The range of mobility varies with the type of boom, but a modern full size boom allows the microphone to be racked in and out, swung from side to side and tilted up and down, thereby covering sound within an arc of almost 180° in the horizontal plane and almost 30° in the vertical plane with a radius between about 4 and 7 m (13–23 ft), assuming the artist is about 1 m (3 ft) in front of the microphone.

The microphone boom can also be tracked or crabbed, i.e. moved sideways with its wheels in parallel track, so that its range of coverage is limited only by practical considerations. The microphone can be turned through about 300° horizontally and about 90° in the vertical plane, and all these movements can be accomplished simultaneously so that changes in direction of sound propagation can also be accommodated.

The boom has the advantage of being controlled by an operator who is very well placed to view the action and to avoid as far as possible creating problems for other aspects of the production. The boom can be raised to clear the picture for long-shots and dropped in for close-ups so that the dual requirements of keeping the microphone out of the picture and matching visual and aural perspective are achieved in the same operation.

Disadvantages

Problems arise in the use of booms:
1. When artists work close to the walls of the set.
2. Where the lighting is in line with the camera shots (see page 88).
3. Where there are pillars or arches or other tall objects in the foreground.
4. When the nature of the programme material or the need to achieve separation requires a close microphone technique regardless of the headroom in the picture.
5. When longshots, or shots with excessive headroom, alternate rapidly so that there is insufficient time to adjust the boom.

Stereo operation

Where stereo is involved, especially for drama, the boom, fitted with a stereo microphone, would normally provide the basic sound pattern, with the output of additional fixed microphones 'panned' into corresponding locations for effects and incidental sounds (see page 70).

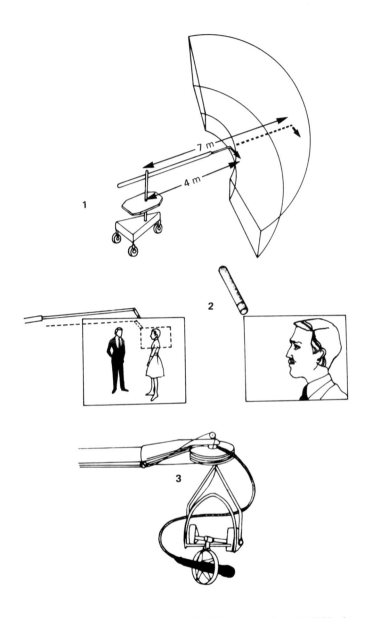

1. A full sized boom is able to cover sound within an arc of about 180° in the horizontal plane and 30° in the vertical direction with a range of radius between 4 and 7 metres, assuming the artist is about one metre in front of the microphone.

2. Held above the action the boom can be quickly raised or lowered to suit the headroom in the camera shot and to match perspective.

3. Detail of boom microphone suspension. Stereo microphones are normally side-sensitive and would be mounted vertically.

Boom operating requires a thorough knowledge of microphone technique and characteristics, visual imagination and a good memory.

Boom Operating

The boom operator's task is to maintain the microphone in as nearly as possible the ideal position and pointing in the optimum direction throughout the action, while causing the minimum restriction and inconvenience to the other aspects of the production.

The boom operator is able to receive directions via headphones from the sound supervisor, but should not have to rely on that. He should be able to assess for himself as accurately as possible the relative levels of the sounds he is picking up and the requirement in terms of aural characteristic (perspective etc.) to match with the camera angles and apparent viewing distances from instant to instant.

Operator's responsibilities

The boom operator must have a keen programme sense and the ability to memorise dialogue. Although he can be provided with a script and shot-list which he clips on a special board, he will not be able to read it while he is involved in a sequence because to look away from the boom would probably result in the microphone coming into shot and could even be dangerous, for example if a seated actor suddenly rose to his feet while the microphone was above his head.

The boom operator must assess the position and direction to point his microphone not only for optimum sound pick-up but also from the point of view of providing maximum discrimination against unwanted sources of sound, i.e. orchestral accompaniment (unwanted on the boom) or PA loudspeakers, etc. To this end he may be more concerned with pointing the dead side of the microphone towards the unwanted source than in pointing the live side directly at the artist.

The boom operator is provided with a reverse talk-back microphone with which, by pressing a key, he can communicate with the sound supervisor. In this way he is able to keep the supervisor informed of the situation and problems that arise on the studio floor, especially during rehearsal, which his commanding position makes him particularly well favoured to observe.

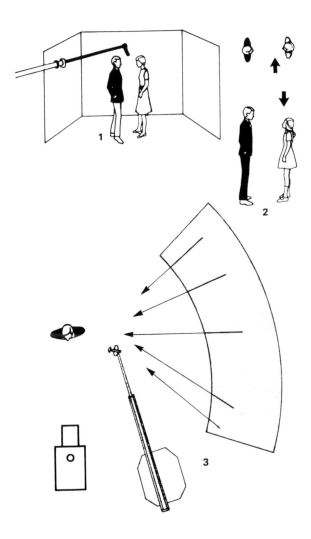

1. The usual position for the boom in a dialogue situation is above the level of the actors' heads and slightly downstage of the action (i.e. toward the cameras). The actual height and distance depend on the camera shots and perspective requirements or the need to discriminate against unwanted sound.

2. The microphone should be positioned and angled to favour the weakest voice or shortest person. The figure gives an indication of the position in plan and elevation.

3. It may be more important for the boom operator to point the dead side of the boom toward the main source of unwanted sound than directly toward the artist.

Television sound operations and lighting are interrelated and require mutual understanding.

Basic Lighting Requirements

One of the most difficult problems associated with boom operating is to avoid making shadows, either on the artists or on the scenery, which can be seen in the picture. Boom operators should be familiar with the basic principles of production lighting and be able to recognise the various light sources and assess the risks involved.

Lamp functions

Very many lamps are employed in most television productions, mainly because of the need to light for a number of different areas and camera angles simultaneously. If the situation in any one position is analysed it will be found that four, or possibly five, basic arrangements are involved.
1. *Key light.* A hard (i.e. point) of light from a lens spotlight intended to provide modelling and to show texture and contour by creating shadows.
2. *Filler.* A soft (diffuse) source of fairly large area, usually obtained by directing a number of light sources through a diffuser or by reflecting light from point sources, which are not directly visible, via a large reflector. These sources are intended to provide light in the shadow areas left by the key light, to show detail and give artistic balance.
3. *Back light.* A light shone from above and behind the artist to put shine on the hair and rim the outline to promote a three-dimensional effect.
4. *Effects light.* Lighting for the set or backing to the artists. It is intended to bring out the architectural features and promote a sense of realism (e.g. sunlight streaming through a window or lamplight at night). Modern light entertainment colour productions often rely upon the use of colour lighting on a white set to effect changes of scene or mood.
5. *Camera headlight.* A light sometimes attached to the camera to soften the shadows under the eyebrows and chin and create catchlights in the eyes in close-ups.

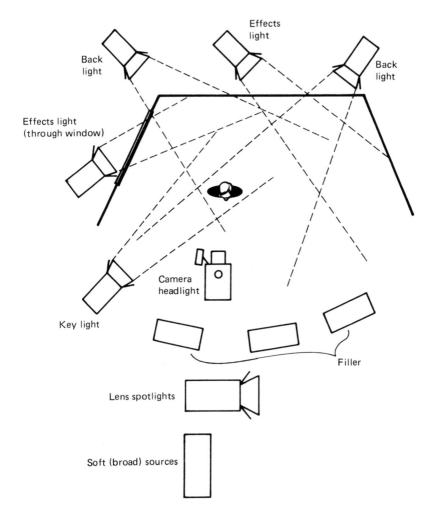

Effects light

Back light

Back light

Effects light (through window)

Camera headlight

Key light

Filler

Lens spotlights

Soft (broad) sources

A typical set in plan view showing basic lighting arrangements. Many more lamps may be required if a number of acting areas and camera viewpoints are involved, but the basic lighting pattern in each position will be much the same.

The Boom and Lighting

The preceding section shows that most shadow problems are caused by the key lights. As the key light is a point source, any object such as a boom microphone placed in its path will throw a distinct shadow. For good portraiture the key light is required to subtend only a small angle to the direction of the artist's face.

If the boom and key light are in line with each other a shadow appears on the artist. The boom must therefore approach from the opposite side of the artist to the key light. When seen in long-shot the artist must work sufficiently far from the scenery so that the shadow of the boom is lost on the floor or on some part of the set that is not in shot. Luckily most situations involve dialogue between two or more people facing each other across the set. In this case the key lighting can come from the sides or even to the rear of the set so that shadow problems are reduced.

Soloists in light entertainment usually require frontal key lights but tend to work with hand-held or stand microphones.

Multi key lights
Continuous-take production can involve the use of a number of acting areas and therefore a number of key lights on each set. Boom operators should study the lighting arrangement before positioning the boom, and lighting directors must take account of boom position before designing their lighting plot.

Heat problems
Sound operators should bear in mind that many of the lamps used in television radiate a considerable amount of heat as well as light. They should be careful not to position microphones, especially electrostatic ones, in close proximity to the lights.

Lamp sing
Light sources that are controlled by electronic dimmers (e.g. thyristors) sometimes radiate quite a loud high-pitched sound due to movement of their filaments. This can be accentuated in certain types of housing such as 'scoops' (a bowl-shaped soft source), which tend to have a marked focusing effect.

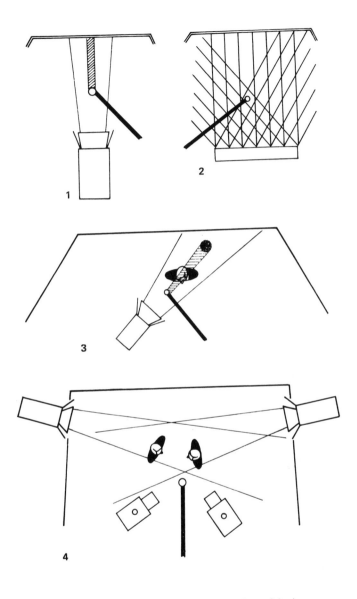

1. A point source of light throws a sharply defined shadow of the boom microphone on to the set.

2. A broad source of light has many overlapping beams in which the individual shadows are cancelled out.

3. When the artist is positioned well forward from the set the shadow of the microphone is thrown on the floor where it cannot be seen.

4. A method of using three-quarter back lights for cross-keying a dialogue sequence. As there is no frontal key light there is no shadow for the boom.

The Boom and Camera Movement

It is only natural for television directors, whose main preoccupation is with directing the cameras, to look upon the microphone boom as a necessary nuisance and to be somewhat apprehensive if it is suggested at a planning meeting that several are required even on one set. In fact modern booms are able to track or crab and are almost as manoeuvrable as cameras. They also have a considerable reach and there is seldom much danger of their restricting camera movement.

Cable arrangements

The essence of the operation is proper planning in relation to the cameras and a well organised system of cabling. The best way to provide flexibility of movement for the boom and at the same time prevent the boom cables from becoming entangled with the camera cables is usually to sling the boom cables from above, preferably by means of special suspension points which can be lowered from the roof grid of the studio. These can consist of simple slung cables and connection position, or sophisticated devices consisting of motor-operated telescopic tubes with connection points mounted at their base.

The connections available should include the microphone lead (possibly with 'phantom' power connection for powered microphones), talkback and reverse talkback lines, foldback circuits (for use when loudspeakers are to be mounted on the boom) and a video cable to enable the boom operator to be provided with a monitor. This can be of considerable assistance where wide variations of picture angle are involved or the boom operator is not able to see the cameras clearly because of some obstruction.

In a large studio four slinging positions would be required, arranged approximately as suggested in the diagram. At the boom end the cable should pass through a 'flyaway' device normally consisting of a metal spiral into which it can be threaded without disconnection unless the suspension system is fitted with a cable counterweight. A few metres of the cable is normally coiled up on the platform of the boom to allow for travel, the slack being taken up and re-coiled as the boom approaches its suspension point.

Boom positioning

The normal position for a boom to work on a set is between, but slightly behind, the cameras. If the cameras are mounted on pedestals it is possible for the boom to work over the top, but if any sort of crane is involved, and in particular a swinging crane, the boom has to work to the side (possibly even to the side of the picture frame).

Some of the largest cranes are provided with spigots to accept microphone boom arms so that the operator can ride with the camera.

90

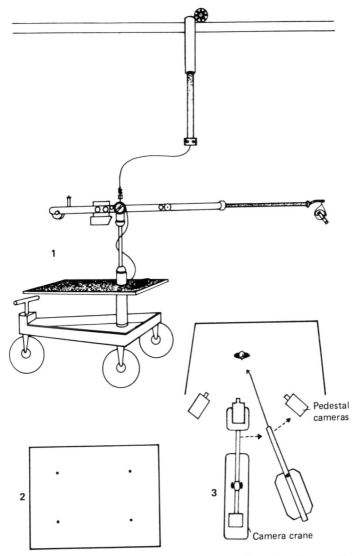

1. Electrically operated telescopic hoist for suspending microphone cable termination from lighting grid.

2. Plan view of large studio showing the general disposition of four boom termination hoists.

3. Boom working alongside cameras, one of which is a crane with a jib that can swing from side to side and up and down. If the crane has to swing toward the boom while elevated, the boom will have to swing also. This is a point that should be taken into consideration at the planning stage as it might require the boom to be repositioned or to have another one on the other side of the crane to take over. If the boom has to execute a wide swing, the 'pram' must be angled so that the operator does not 'run out of platform'.

Planning for the Boom

Planning for the boom centres around two types of scale drawing, the studio plan and the elevations. Sometimes models are also used.

The studio plan
The studio plan consists of a 1:50 scale plan of the studio, usually marked out in 6 mm (approx. ¼ in) squares, each to represent 30 cm (1 ft). The outline of scenery is marked on the plan to scale, and the starting positions for the cameras and booms are marked by means of a special stencil. The positioning of the cameras and booms is decided at a planning meeting prior to the production day, attended by the director, sound supervisor, lighting director, set designer and any other specialists concerned at that stage in the proceedings. It is explained that the marked position can give only an approximate indication of the starting position because both cameras and booms are mobile. No matter how careful the planning, shots develop and positions become modified as a result of rehearsal experience. Nevertheless it is necessary to have a pretty firm basis from which to work.

During the course of a production the cameras and booms will probably be required to work in a variety of different areas. These are identified on the plan by the camera numbers or boom letters and position numbers (cam. 1 pos. A; boom A pos. 2, etc.).

Part of the art of camera planning concerns arranging the cabling so that the cables do not get tied up in knots as the cameras move about, and this goes for the booms too if the cables are not slung.

The elevations
The elevations are detail drawings of the vertical appearance of the sets. This can be very useful at the planning stage in judging heights, particularly if obstacles are to be encountered such as roofs, pillars, arches etc.

When very complicated structures are involved it is usual to make models of the sets from balsa wood or cardboard to the same scale as the plan. These can be invaluable in helping to assess the optimum position for microphone coverage and the hazards of boom operations.

The actual detail of boom planning is bound up with the optimum sound requirements. This depends to a large extent on the nature of the production and is dealt with in detail in later sections.

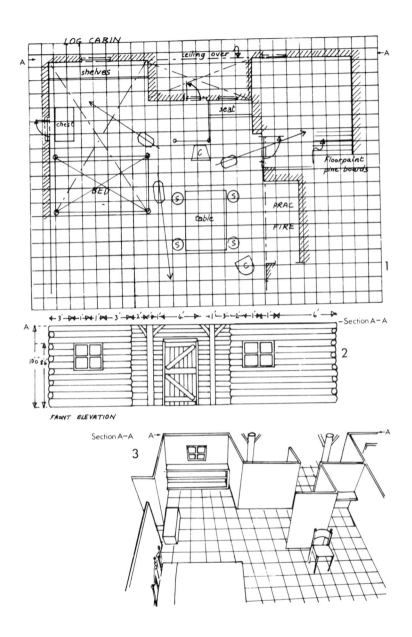

Labels within figure 1 (plan): LOG CABIN, ceiling over, shelves, seat, chest, BED, table, Floorpaint pine boards, PRAC FIRE, 1

Figure 2: Section A–A, 2, FRONT ELEVATION, 10'0", 8'6"

Figure 3: Section A–A, 3

1. Marked-up studio floor plan, showing disposition of booms.

2. Elevation, indicating the vertical appearance of the sets.

3. Model of complex item of scenery.

Production Talkback Arrangements

Television production is, above all, a team operation which relies heavily on good communications. Continuous-take or live production would not be possible unless all the operators concerned were able to hear the director simultaneously. For this reason, all camera and sound staff, and any other operators working on the studio floor who are required to take cues, wear headphones (often called 'cans' because of the can-like appearance of the metal ones used for many years in radio and television).

Director's talkback

The main talkback source is the director's (or producer's) microphone. This is normally live throughout the performance. It is fed to all the operators working on the studio floor as mentioned above, and also to any other areas concerned with the production. These include the lighting and sound control rooms (if separate from the production control room), any telecine and videotape channels associated with the production, any remote sources that may be involved, and even loudspeakers in the artists' dressing rooms.

The director's talkback must be clear and of reasonably consistent volume, so directional microphones are usually employed (to discriminate against the programme loudspeakers and other ambient sounds) and a volume limiter is incorporated. The director is also provided with loudspeaker talkback into the studio, controlled by a key, but seldom uses it because it tends to cause confusion.

Floor manager's talkback

The director's directions are relayed to the artists by the floor manager, who is therefore required to be very mobile. He normally hears the director by means of a low power (about 500 mW) transmitter and a small receiver which he carries in his pocket or a sling harness. A reverse talkback transmitter can be provided for the floor manager to speak back to the producer, but this facility is not often required as there are usually programme microphones, e.g. a boom available in the studio.

Camera talkback

The director's talkback is fed to all the cameramen's headphones via the camera cable. The cameramen have the facility to listen also to programme sound. This can be mixed with the director's talkback to the required proportion to augment that inevitably picked up from the loudspeakers in the production control room. In some systems the headphones are split to provide talkback in one ear and programme in the other.

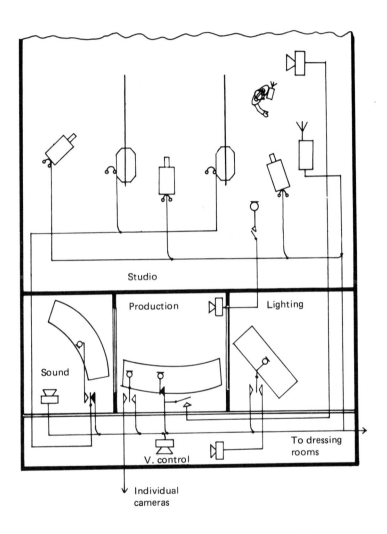

Studio

Production   Lighting

Sound

To dressing
rooms

V. control

Individual
cameras

Typical studio talkback arrangement. The producer's talkback is the only circuit that is normally connected to everybody continuously. That of the sound, lighting, technical manager and vision control is directed to the respective personnel via switches.

Sound and Lighting Talkback

Just as the director uses his talkback to direct the movement of cameras about the studio floor, so the sound supervisor and lighting director direct the sound floor staff and electricians over separate talkback systems.

Sound talkback
It is not possible for the operators at the microphone to judge the effect of their work. They need guidance during rehearsal, although good boom operators cultivate such a good memory that they should manage with little direction during transmission. Nevertheless they need to be told when changeovers between the boom have been completed and they are released to go to their next position. As they seldom have time to read a script it is a good idea to warn them of important cues and moves.

The sound operators on the studio floor hear the same talkback as the cameramen until the sound supervisor wishes to speak to them. The sound supervisor then operates a key, a foot switch, which exchanges his microphone output for the director's. This also gives the sound operators a short burst of tone as an indication that the sound supervisor is intercepting the circuit.

Reverse talkback
The boom operator is provided with a reverse talkback system consisting of a key-operated microphone mounted on the boom. He uses this to warn the sound supervisor of any problems that arise in the studio that he may not be able to see, and to cue him as to the best moment to change over from the output of one boom to another in a take-over situation.

Lighting talkback
A similar arrangement also exists for the vision control staff and the lighting electricians. They normally listen to the director's talkback until it is intercepted by the lighting director. Lighting directors normally have available small two-way radios working between lighting control room and studio floor to facilitate lamp adjustment during setting periods and rehearsal.

Wall points
All three talkback circuits (director, sound and vision) are available at all wall points and suspended boom connections.

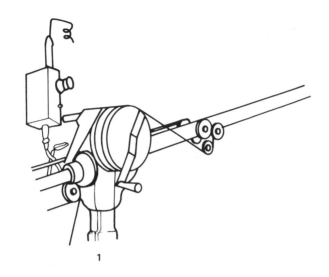

1

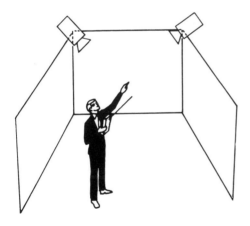

2

1. The boom operator can be provided with a microphone and switch box mounted on the boom. If he wishes to talk to the sound supervisor during the action he has to approach very close to the microphone and use a very quiet voice. Alternatively he can be provided with headphones which incorporate a small close talking microphone.

2. A typical walkie-talkie communication set in use for setting lamps.

Sound Control

Most reproducing equipment has a volume control, so neither the artist nor the sound operator can have any say over the actual volume at which their material is finally reproduced. They can merely affect the *relative* loudness of the various parts of the programme and its individual elements.

Basic requirements
There are three basic requirements:
1. To keep within the dynamic range of the system. In most cases, such as recording and live transmission, the upper limit is set by the onset of overload distortion and the lower by the threshold of intrusive noise.
2. To adjust the scale of volume range to suit the listening conditions and environment. Most types of programme material, from music to drama, involve a range of volume that would be quite unacceptable in average domestic circumstances.
3. To ensure that items produced in different areas and at different times match each other and in broadcasting maintain a smooth balance between successive programmes.

Thus there are two aspects of volume — the technical and the artistic. As we are not able to measure absolute sound levels with our ears and have a poor memory even for relative levels, we need a meter (as described on page 100) and a set of rules regarding the proper level for each type of programme material if we are to keep within the technical limits and provide a balance between successive items or programmes. But the meter does not help us to make artistic judgments or even assess loudness, which is largely subjective.

Use of loudspeakers
The only way to effective programme control is by means of a good quality loudspeaker, making use of a meter to 'calibrate' the ear in the way that a car driver judges his speed without watching the speedometer all the time.

The importance of listening levels
Practically all operators professionally engaged in sound control listen at very high volume. This is necessary to achieve the right degree of concentration and because otherwise there is a tendency to 'wind it up' out of sheer enthusiasm. Unfortunately this can give a very unreal picture of what the listener will hear and can lead to a misleading balance, particularly as between vocal and accompaniment or speech and sound effects. It is therefore advisable to 'dim' the loudspeaker from time to time (perhaps occasionally listening on a small loudspeaker of the type fitted in the average television receiver) and get used to making allowances for the difference.

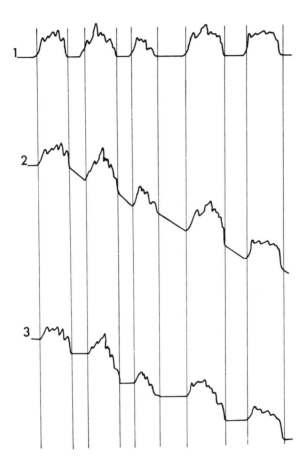

1. *Speech modulation.* Most sources of sound are continually varying and interrupted in character and a fade which continues through the gaps in the material would result in a succession of steps in the output.

2. *Unobtrusive control.* If the intention is for the operation to be as inconspicuous as possible, which is the case for most types of material and particularly music, the action should occur in the gaps in the material or during rapid fluctuation of level.

3. *Control for dramatic effect.* If it is intended to make an obvious smooth fade for dramatic effect (e.g. a slow fade on dialogue is the conventional method of suggesting a change of scene in radio drama) movement of the control must occur only during the material, preferably during sustained phrases.

Programme Meters

It is necessary to keep programme volume within prescribed limits, both for practical and artistic reasons. Although the proper way to control programme volume is by listening, it is necessary to have meters to set volume levels and assist in the assessment of volume range, i.e. to calibrate the ear. There are two basic types of programme meter in professional use.

The VU meter

The Volume Unit meter, in its simplest form, can consist of a moving coil meter and rectifier. The meter has two scales, one above the other. One, calibrated in decibels, is intended for use with steady tone for line-up. The other is calibrated in volume units, numbered in black from 0 to 100 over about two-thirds of its range, with the last part uncalibrated and marked in red. This scale represents percentage of full modulation and is intended for use on programme.

The meters can be made to give an accurate indication of zero level (0 VU) on steady tone, but the way they react to programme material depends upon the ballistics of the meter (rise time and overshoot) and the circuitry. Meters used for professional purposes, where levels have to be set and compared, should conform to the American Standards Association specification. Even these will give different, but matching, readings with different material. Due to the rise time of the meter, impulsive sounds starting from a low level will read lower than their actual level; sustained sounds tend to read higher and fluctuations near the top of the scale can overshoot and read high. For this reason VU meters have to be interpreted from experience rather than just read. Line-up level is usually taken to be +4 dBU (1.73 volts) to read 0 VU for an actual peak level of +8 dBU on average material.

The peak programme meter

A more accurate visual indication of volume level is given by the peak programme meter. The PPM has a rapid rise time (2.5 ms) and a slow recovery time (about 3 seconds) actuated by charging a capacitor and discharging it through a high resistance. This action makes the meter easy to read and gives a reasonably accurate indication of peaks, which is often the most important consideration. The meter is calibrated in decibels with figures from 1 to 7. There is also a zero mark and a full-scale deflection mark. Normal practice is to line up with tone at some middle frequency (such as 1 kHz) to a level of 8 dB below the maximum — represented by the figure 4 on the PPM.

1

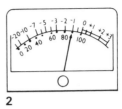

2

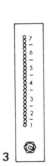

3

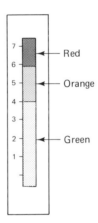

Red

Orange

Green

4

1. *Peak programme meter.* The figures are white on black to reduce eye strain. The law is quasi-logarithmic so that the scale is reasonably evenly spaced, with 4 dB between each of figures 2–7. Line-up level is usually set to 4 and maximum level (+8 dBm) is 6. The extra 4 dB to 7 is to give an indication of the degree of overmodulation.

2. *VU meter.* There are two scales. The upper one is marked in decibels (red above the zero mark, which represents full modulation). The lower scale is marked 0–100 and represents the percentage of full modulation. The scale is uneven, being largely cramped towards the top end.

3. *Visual display meter* suitable for use in individual channels where space is limited. This can consist of a stack of light-emitting diodes. Different colours can be used to represent low, normal and overload levels.

4. *Plasma bar graph.* Various colours are used to indicate low, normal and high levels. This type of meter can usually be switched to display PPM or VU readings or to freeze peak values.

The Points of Control

On page 98 we discussed the need for control. Now we consider the points at which programme control is exercised.

Professional sound control desks come in all sizes and degrees of complexity to handle any number of channels from less than six to more than forty. On the larger control desks, programme control can usually be exercised at various alternative positions in the chain.

Pre-set (or 'balance') attenuator

For television sound the range of levels applied to the channels can be enormous, varying perhaps from whispered dialogue picked up by a microphone at an excessive distance (in order to be out of shot) to a shout or a musical instrument at very close range. This can represent a variation in microphone level possibly of up to 60 dB (i.e. 1 000 000:1). In order to enable the apparatus to cope with such a range, television desks are usually equipped with pre-set controls in addition to the channel controls. These are usually small variable attenuators by which the gain of each channel can be adjusted.

Channel fader

The whole essence of effective programme control is that it should affect only those elements of the sound that require adjustment (imagine the sound of a whole orchestra being diminished to accommodate a beat of a drum). Thus the most important controls are usually the individual channel faders. For stereo operation 'pan-pots' can be incorporated at this stage.

Group fader

Although ideally control should be exercised on the individual channels there are circumstances when the operator has no time, or insufficient fingers, to adjust all the required number of controls simultaneously. It is usually possible to select a number of sources that can be conveniently grouped together. These might include, for example, several microphones in use for a large choir, or a section of an orchestra, or even the whole orchestra; the object is to group together sources that might have to be balanced as a section against other sources without disturbing their internal balance.

Main fader

Last in the sound desk comes the main fader (or main faders, if more than one output is provided). The main function of this control in modern practice is to enable a balanced mix of all required sources to be faded in and out simultaneously and to set a convenient overall level. For this reason it is seldom altered throughout the performance and can be considered almost a pre-set device.

SOUND CONTROL DESK

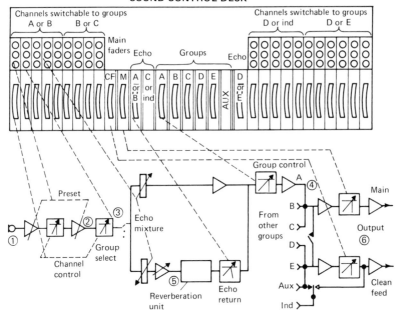

Basic sound control desk, omitting equalisation and ancillary features for the sake of simplicity. A stereo-capable desk would provide two main outputs. The channels would be grouped in pairs with 'pan-pots' between them.

Typical programme chain of modern sound mixing desk. Average levels at various control points are:

1. Input — 70 dB.

2. Pre-set amplifier controls (set to 10 dB below full gain) — 18 dB.

3. Channel fader (set to 10 dB attenuation) — 18 dB.

4. Group control (set to 10 dB attenuation) — 18 dB.

5. Input to reverberation unit (at 50/50 mixture) — 18 dB.

6. Output (with main control set to 4 dB attenuation) zero level.

The maximum gain of the desk with all controls flat out is approx 105 dB. Gain with controls in the above positions for optimum signal-to-noise ratio and minimum risk of overload is 70 dB.

Compressors enable higher average sound levels to be maintained without overload.

Automatic Control: Limiters and Compressors

No automatic device can provide the necessary aesthetic judgment and finesse for effective programme sound control. Nevertheless automatic control systems (known as compressors and limiters) have considerable application in sound operational practice to assist in the 'tamping' of wildly varying and unpredictable sources and as a long stop to protect equipment from possible overmodulation.

The limiter

The limiter is a device through which programmes can be passed without either gain or alteration of the signal until a critical value is reached. If the input signal rises above this value (usually called the onset point) the gain of the system is automatically reduced (below unity) to such an extent that the output is not allowed to rise significantly above the limiting value. This limiting action is caused by reduction of the amplifier gain, not merely by cutting off the peaks of the waveform which would be 'peak chopping' and would result in very heavy distortion. Limiters are commonly employed at transmitters to protect the circuits from overload and, in the case of frequency-modulated transmitters, to prevent over-deviation into adjacent channels.

The compressor

The compressor is similar to the limiter, in that above the onset point the gain of the system is reduced, but less dramatically, so that an increase in input level above this value produces an increase in the output but to a lesser degree. The extent of the gain reduction is usually adjustable and is called the *compression ratio.* Other controls usually available on compressors are:

1. The *onset* or *threshold point* (the value at which the action commences).
2. The *attack time*, being the time required for the action to take effect following a sudden increase in the level.
3. The *recovery time*, which is approximately the time required for the gain to be restored to unity after the high-level signal has been removed.

The attack time is set as a compromise between too fast, which can alter the shape of the waveform and thereby cause distortion, and too slow with consequent overshoot.

Recovery time is usually set to make the variations of gain as inconspicuous as possible without allowing isolated peaks to depress the level for too long.

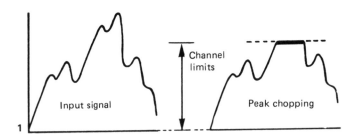

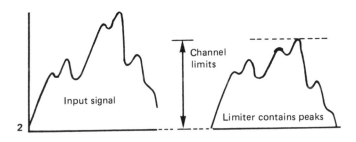

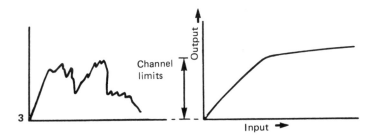

1. Effect of peak chopping with severe distortion.

2. A limiter should cause gain reduction to contain the peak of the waveform.

3. Input/output curve for a limiter showing the 'knee' or onset point of compression.

Sound Control Desks

Sound desks come in all sorts of sizes and complexities. The keynote is to select a desk with all the required facilities and a large margin of extras, with the flexibility to meet future requirements (which always seem to grow).

Some desks allow for extra channels to be added to the main busbars as requirements increase, and can have extra channels cascaded, their output being brought back as input to a regular channel, which then becomes their group control.

Most desks have provision for the channels to be used as either microphone or high-level inputs. It is useful to have provision to feed +48 V DC down the microphone line to polarise condenser mics or supply DI boxes.

There must be a comprehensive cueing system whereby various items from the input can be 'folded-back' to specific monitor loudspeakers in the studio, and this can include selective talk-back.

If there is likely to be a requirement for stereo all channels should be provided with pan pots to divide their output between the L and R outputs as required.

Comprehensive monitoring on the control room loudspeaker is required, selectable to individual channels and groups, both before and after the faders.

EQ and reverb

Several sources of reverb should be available, assignable to individual sources or groups, as different characters of reverb are required for different parts of the balance.

Equalization and a means for inserting volume compressors should be provided in each channel, with provision for more complex adjustment (graphic equalizers etc.) at the group stage.

Auxiliary circuits

A number of auxiliary outputs are necessary which can be fed from individual channels or groups isolated from the main output to provide mixed-minus (clean feeds) to contributing areas, such as remotes, which need to have an output that does not include their own. Aux outputs and return can be used for multi-track recording and for interposing reverb and other devices.

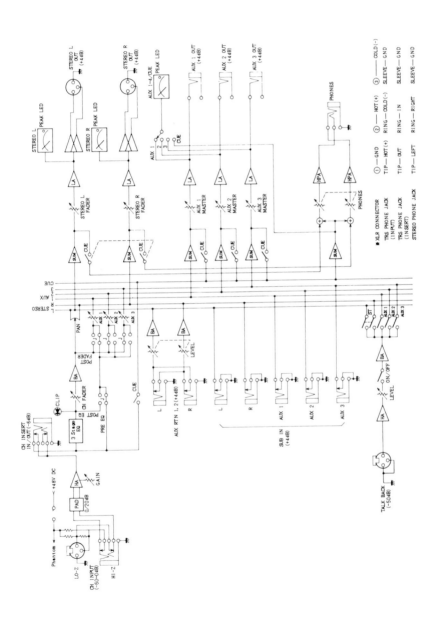

Block diagram of a basic sound mixing desk, showing one input channel.

107

Automated Sound Mixing

Where complex music sound mixing is involved and the programme is not live, there is much to be gained by pre-recording, especially if multi-tracking can be used.

Multi-tracking technique

The complexity of modern balance technique is such that it is almost impossible to obtain the best results by mixing the sounds as they are made. It is much better to record each element on separate tracks of a multi-track recorder so that each can be individually treated and, if necessary, repeated without affecting the rest. Moreover, most of the treatment and manipulation of the sound can go on after the musicians have left the studio. However, this 'mix down' process can be tedious and needs to be efficiently handled if optimum results are to be obtained without extravagant use of studio time. This has led to the introduction of *automated mix-down.*

Automated mix-down

During the first attempt at mix-down a signal is recorded on one track of the multi-track tape which, in one system, is coded to represent the variations of the individual channel faders in relation to position along the tape. Alternatively a time code (usually the SMPTE time code) can be recorded and used to synchronise a memory system for the same purpose. Once the fader settings have been memorised on the tape, the mix can be repeated and the channel faders modified as required (and the memory updated) until an optimum master mix is obtained. The mix information can be stored on floppy disc.

Mechanical fader control

The method by which the fader settings are reproduced can be mechanical or electronic. In the mechanical system the faders are physically moved by servo motors and can be seen to be operating as though by an invisible human hand. Touching the fader knob has the effect of disabling the fader servo so that the settings can be modified and re-memorised by this natural and instinctive operation.

Electronic channel control

In an alternative method the *effect* of the faders, not their position, is automated. This technique is facilitated by the use of voltage-controlled amplifiers (VCA). These amplifiers have their gain controlled by a DC voltage, typically in the range $-2V$ to $+10V$ DC. The control voltage adjusts the amplifier gain logarithmically and is usually arranged so that an increase of one volt represents an attenuation of 10 dB. VCAs are usually set up so that zero volts from the control fader produces zero gain, $+10V$ gives 100 dB attenuation (cut off) and $-2V$ gives 20 dB gain. This allows for 20 dB 'headroom' above normal settings.

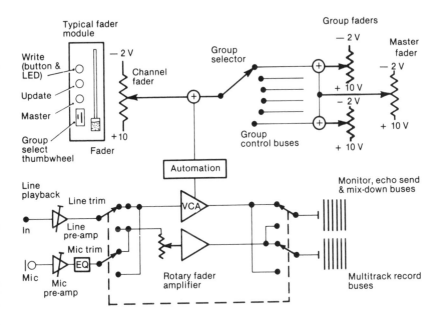

The control voltage for the VCAs is varied by the channel faders. It can also be supplied from group faders by selecting the appropriate bus bar to add to the fader control voltage. Finally the group faders can be controlled by master faders by simply adding their voltage to the groups, the actual volume adjustment still taking place in the channel faders.

One advantage of this system is that the individual channels can be controlled as groups without their outputs being mixed so that they can be recorded completely separately on the multi-track tape.

To make changes in the mix on subsequent attempts it is necessary to find the position on the fader that matches the VCA control voltage. This is shown by two LEDs, one of which lights up when the fader is too high and the other when it is too low. When both light, the fader setting is matched; the 'write' button can be pressed and the fader moved to a new position to update the memory.

Some mixing consoles have alternative paths (controlled by rotary faders) for making the original recording.

109

Sound Shaping

No matter how good the audio equipment in use, there will be many occasions when the sound can be improved by using equalization and various 'sound sweetening' devices, either to tackle the problems caused by microphone technique imposed by vision considerations or for artistic effect.

Equalization

Most sound control desks incorporate equalizers for shaping the frequency response of individual channels. These can be used to equalize (as the name suggests) the output to produce a flat response, or to enhance some aspect of it.

In considering the use of eq. it is worth remembering the characteristics of our hearing. Our ears are most sensitive to frequencies in the 3–5 kHz region, so an increase in this area tends to give greater *impact* to the sound, i.e. more of an edge to cut through other sounds. Also this is the area where the sibilants in speech and vocal sounds are most prominent. It gives clarity of diction and a sense of *presence* or closeness. This is the purpose of the so-called *presence filters* (usually parametric equalizers with variable width (Q) and height peaks that can be moved up or down the frequency scale to suit the voice). In general we tend to associate closeness with the ability to hear the high frequencies because these are the more attenuated with distance by friction with the air. On the other hand too much HF boost, particularly in the 3–5 kHz region, can give the sound an unpleasant harsh quality which soon becomes tiresome to the listener. Bass boost can give the sound added depth and sonority, but too much can make it dull and put a strain on the reproducing ability of the average domestic receiver, resulting in some unpleasant noises.

Equalizers

These are necessary to correct for deficiencies in audio response. For example, a clip-on microphone used under clothing could be lacking in HF and 'presence'. The output could be compared with a regular mic in a good position, say in a boom, and the response adjusted with a *graphic equalizer*, which gives separate frequency response control of each octave, to match the sound.

Other correcting equalizers include *shelving filters*, which provide a flat response for most of the range with a rapid cut-off at the top or bottom, useful for tackling spurious noises like ventilation noise and for producing telephone and intercom effects. *Notch filters* provide very steep dips in the response at specific frequencies and are useful for removing mains hum or resonances.

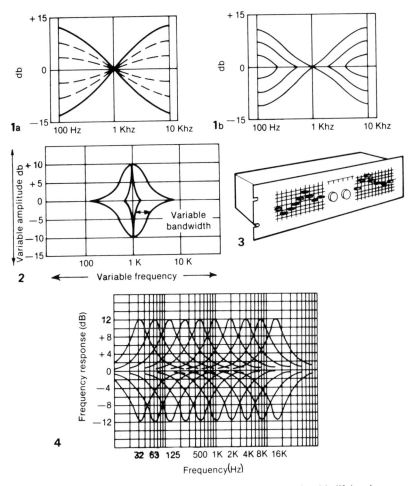

1. *Response selection amplifiers.* (a) In one type the bass and treble lift/cut is adjusted by varying the slope of the response curves about a fixed middle point, in this case 1 kHz. (b) In another type the bass and treble lift/cut is altered by varying the turnover frequencies.

2. *Parametric equalisers.* The bandwidth of the peak or trough can be adjusted between about 0.1 and 5 octaves. Its amplitude can be adjusted between zero and about ±15 dB and its position slid along the frequency scale. The width of the peak or trough (Q factor) can also be adjusted.

3. *A typical stereo graphic equaliser.* This type of equaliser gives control to ±12 dB over 10 octaves centred on 32, 63, 125, 250, 500, 1 k, 2 k, 4 k, 8 k and 16 kHz. The positions of the sliders give a 'graphic' indication of the shape of the response curve.

4. *Graphic equaliser curves.* Careful design is necessary to prevent phasing distortion due to interaction between the overlapping peaks. Most graphic equalisers produce a response that looks very distorted when viewed on an oscilloscope but with good design this is not audible.

Artificial Reverberation (Echo or 'Reverb')

Reverberation (see page 24) is essential in varying degrees for all types of programme material. Where there is too little natural reverberation, or close-microphone technique is used, it can be added artificially.

Echo room

The simplest way to obtain artificial reverberation is to place a loudspeaker and microphone in a room with reflective walls. A proportion of the output of the respective microphone is fed into the loudspeaker in the reverberation chamber (echo room) and then a proportion of the reverberant sound picked up on the microphone is added back into the main output.

Reverberation plate

A convenient method of obtaining artificial reverberation is the reverberation plate. It consists of a steel sheet approximately 2.5 m × 1.5 m (8 × 5 ft) suspended vertically on springs. A moving coil transducer is fixed at a critical point near the centre and a contact microphone is mounted near one edge (two for stereophony). When a signal is applied to the transducer a series of very complex vibrations radiate out to the sides and are reflected to and fro between the edges until they die out in the manner of reverberation.

The reverberation time can be adjusted by altering the spacing between the plate and another plate of similar size made of porous material held parallel to the steel plate. The porous plate damps the reflections in the steel plate. The adjustment can be remotely controlled.

Reverberation springs

A relatively cheap and compact echo device consists of two sets of springs of different lengths joined together at the ends. Sound waves are induced into one end of the springs in a torsional manner and received at the other. The energy reflects backward and forward between the two ends until it dies away, giving a reverberant effect. The reverb time can be adjusted between 2 and 4.5 seconds by altering the tension of the springs.

Digital reverberation

In a typical digital reverb device the signal is converted to digital format and passed through a pre-reverb delay to a tapped delay line with a series of outputs, each being progressively attenuated and multiplied many times (requiring a very high sampling ratio) to represent the vast number of reflections in normal reverberation and fed back into the input to recirculate. Comb filters with different delays are added in parallel. They cause phasing effects (as in natural reverberation). There is also tone

1. Plan of typical arrangement of a reverberation chamber (echo room). If possible the walls should not be parallel to each other. The baffles create a longer initial path between loudspeaker and microphones and help to break up standing wave reflections.

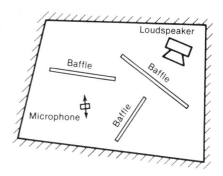

2. Reverberation plate. The specially selected metal sheet is suspended by springs at each corner in a wooden box. Vibrations are induced into it by a moving coil transducer and picked up by one, two (in the case of stereo) or four (for quadrophony) ceramic contact microphones.

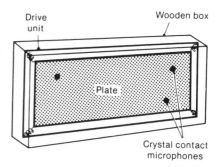

3. Reverberation springs. Usually arranged in pairs, each with two springs of different lengths joined together.

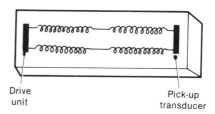

control, because natural reverberation tends to fall off in the high frequencies.

Time delay

Whichever reverberation device is used it is likely to be more natural if the sound is delayed before the reverb. This is because, in the natural situation, the sound waves have to travel to the walls and back to the listener (at about 1 ms per foot) before the first reflections take effect. *Digital delay devices* are available that provide variable delays of up to several seconds and are ideal for delaying the input to electro-acoustic reverberators. *Digital reverberation equipment* usually provides the facility to select various values of pre-delay, reverberation time and diffusion. Computer control of the various reverb and tonal characteristics provides pre-set programs to simulate rooms of different sizes and particular qualities to 'fatten' vocal or instrumental sounds.

113

Using Reverb

Unless the natural acoustic conditions happen to be perfect, most sounds can be improved by starting with too dry an acoustic, or with close microphone technique, and adding artificial reverberation of a suitable character. The various parameters are as follows.

Pre-delay This represents the time that, in the natural situation, the sound reflections would take to build up before reverberation becomes effective. It is related to the size of the room. Early reflections can enhance presence and impact.

Amount of reverb The ratio of direct to reverberant sound (echo mixture), this determines the apparent volume and warmth of the sound. We tend to associate reverberation with sound 'filling the hall'. (If you whisper in a large hall you will not be aware of the reverberation. Shout and it will be very obvious. The reverb time of the room does not change, just the amount of the reverb that is loud enough for you to hear.)

Rate of decay This determines the apparent size of the room. A slow rate of decay suggests a large room (about 2 seconds for a concert hall, 5 seconds for a cathedral). A large proportion of reverberant sound with a rapid decay would suggest a small, very reverberant room (e.g. a bathroom). This type of reverberation can be used to 'fatten' the sound, give the effect of power, particularly in the bass, and cover imperfections. (It is encouraging to sing in the bath.)

Reverberation frequency response Analysis of various studios and concert halls shows that the best music studios have reverb frequency responses that are far from flat. Usually they have a marked, but smooth, increase in the 200 Hz to 2 kHz region. This gives a warmth and sonority to the sound. The frequency responses of reverb devices should be similarly adjusted to suit the character of the sound.

Diffusion Digital reverb devices provide the ability to adjust diffusion, i.e. the smoothness of the decay. Natural acoustics are usually best when the diffusion adopts a smooth expotential curve. However this may not always be ideal, for example if reverb is being used in a particular channel to 'fatten' the sound, say, of a drum. Low diffusion results in a grainy sound which can suit vocals.

Depth The apparent distance of a sound source is a function of pre-delay, ratio of reverb to direct sound and decay slope.

Chorus Digital reverb devices can offer effects such as multiple repeats with slight delayed timing and pitch changes (phase) to have the effect of multiplying the number of musicians and thickening the sound.

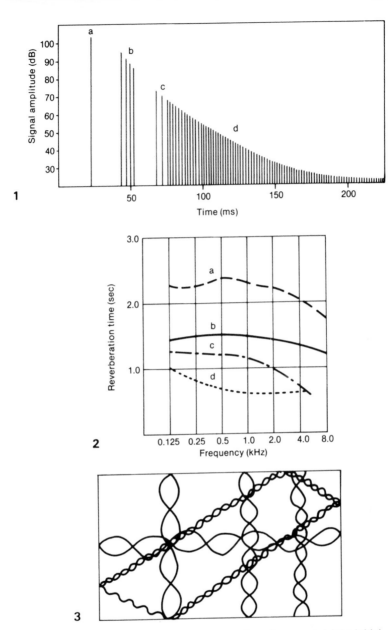

1. Illustrating the response of an auditorium to an impulse sound: (a) the initial sound; (b) the first few reflections from comparatively close reflecting surfaces; (c) reflections from nearest walls; (d) multiple reflections from the body of the hall.

2. Typical reverberation times for various types of room and for various frequencies. (a) Very large hall; (b) concert hall; (c) large lecture room; (d) living room.

3. Some standing wave patterns (eigentones) in a room. These cause resonances and colouration in a small room.

115

Time Delay: the Haas Effect

There is a relationship between the order of precedence and the relative volume in which sounds are heard that determines the direction from which they appear to come. This is an important consideration in all multi-source systems whether they concern stereophony, surround sound or even a simple PA artist reinforcement system.

The Haas effect

Haas investigated the effect on the listener of two sources radiating the same sound from two different directions in the horizontal plane. These could be, for example, two loudspeakers fed with the same signal or a man and a PA loudspeaker reproducing his voice.

If the loudspeaker is nearer to the listener than the man (as often happens with PA systems), the sound from the loudspeaker will arrive first. If the sounds from the two sources have the same volume at the listener's position, his impression will be that all the sound originates only from the loudspeaker and none from the man.

If a time delay of between about 5 and 50 ms is introduced into the feed to the loudspeaker, the impression will be reversed: none of the sound will appear to come from the loudspeaker although this contributes to the volume and reverberation of the sound. If, however, the volume of the delayed sound from the loudspeaker is increased relative to the direct sound from the man, a point will be reached when it overrides the time delay and the sound again appears to come only from the loudspeaker. The relationship between volume level and time delay in establishing the apparent direction of sound is illustrated by the graph opposite.

It will be seen that the maximum effect occurs with delays between about 10 and 30 ms. After about 60 ms transient sounds begin to be discernible as separate echoes. It should be remembered that the speed of sound in air introduces a natural delay of the order of 3 ms/m (1 ms/ft).

Artificial time delay

The ability to insert time delay has enormous advantages, apart from the obvious one in the PA example quoted above. Time delay inserted in artificial reverberation systems improves realism, as it is normal for the first reflections to arrive about 50–100 ms after the direct sound. The ability to make sound appear to come from one direction when the bulk of it is coming from another has considerable artistic possibilities. Time delays can now be provided by digital means.

116

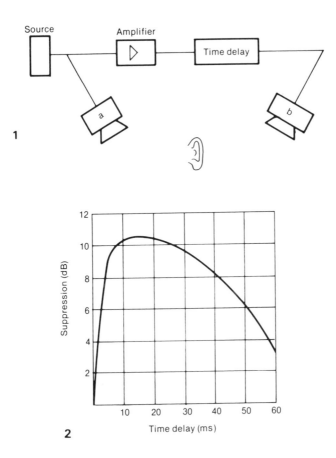

1

2 Time delay (ms)

1. If two similar loudspeakers (a and b) are fed from the same source one through an additional amplifier and variable time delay and listened to from an equal distance the following effects will be noticed: (1) If the loudspeakers produce the same volume, in phase, the sound will appear to come from a point equidistant between them. (2) If b is made louder than a all the sound will appear to come from b. (3) If the output of b is delayed with respect to a but the two volumes are the same, all the sound will appear to come from a. (4) The effect of the time delay can be overcome by increasing the amplification to make b louder than a.

2. The curve shows the relationship between volume and time delay in establishing the apparent position of a sound source. This shows, in the example of 1 above, that if 10 ms delay is inserted in the feed to loudspeaker b it would have to be over 10 dB louder than a to make the sound appear to come from a point between them.

117

Sound can have an enormous volume range. The trick is to preserve the psychological effect without exceeding the technical parameters.

Signal Processing

Programme makers naturally want their programmes to be heard and that means all of it, the quiet bits as well as the loud. Most domestic equipment, especially television sets, is not capable of reproducing the range of sound encountered in real life, and the viewers would not put up with such volume in the domestic situation anyway.

Levelling and compression
Levelling is the long term control of signal level to maintain a relatively constant output level without lessening the impact of short-term variations which have been 'tamed' by the use of *compression*. *Process balance* determines the ratio between levelling and compression.

The 'ducker'
This is a form of compressor in which the output of one signal is controlled by another. It can be used for 'voice-overs' to reduce (duck) the level of music or effects when a narrator or DJ speaks.

The noise gate
When compression is applied to a signal it brings up the gain when the level is low and that includes the noise, which will be obvious in gaps in the programme. To overcome this a noise gate can be used. It can consist of a simple electronic switch which cuts off the output in the absence of signal and restores it when the minimum acceptable level is reached. Unfortunately in many circumstances the sudden switching of noise only tends to draw attention to itself. A more elegant method of noise gating is the *expander*. This is virtually the opposite of the compressor. It has a gain greater than unity up to the threshold point so that it effectively increases the distinction between wanted and unwanted sound.

De-essers
Simple compressors operate on the most powerful component in the frequency range of the programme. In the case of speech it is the lower middle frequencies in most words that drive the compressor, reducing the gain, which can be restored in time to accentuate the weaker consonants (particularly 's' sounds at the ends of words) resulting in an unpleasant sibilance. Many compressors employ de-essing circuits which introduce pre-emphasis into the side chain to equalise the controlling effect.

Multi-band peak limiting
One problem in the use of limiters is the effect of 'pumping' when strong signals of one frequency take control of lower level programme. For example, a bass drum beat could cause the general sound of a band to 'pump' up and down. This can be overcome by dividing the limiting action

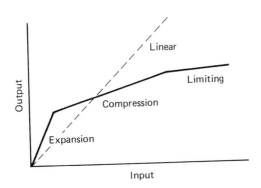

A composite compressor, incorporating an 'expander' to enlarge the distinction between wanted and unwanted sound (a form of noise gate), a compressor to 'tame' widely varying material and a limiter to catch overshoot.

into frequency bands (usually low, middle and high) with separate threshold and slope controls for each. This can also help to solve the de-essing problem.

Companding
This is the process of improving the signal/noise ratio by compressing the volume range of a signal before a process (such as recording or line transmission) and expanding it afterwards. The problem is to make the two functions track accurately throughout the range. (See page 206.)

Talks and Lectures

Talks and discussions tend to be among the simpler productions from the sound point of view, but that is not to say that they are not important. The essence of a talks programme is the imparting of information and it is essential that the speech is clear.

Introductory music
Most talks productions begin with introductory music. Unless the talk happens to be about music it is likely that a proportion of the potential audience will appreciate the talk more than the music. If the music is too loud they will switch off or at least turn down their volume controls to the detriment of the rest of the programme. It is important therefore not to allow the introductory music to be too loud. It should normally peak less than the speech.

Microphone arrangements
For a speaker in a static position a slung microphone can be used overhead if a long-shot involving headroom is not required. Alternatively a boom can be used. If the speaker is seated, a floor-standing microphone is suitable, perhaps placed to the side just out of shot. If the microphone is likely to come into shot, as in an introductory long-shot, an electrostatic type with one-metre extension tube is less obtrusive.

If the speaker has to walk about at some point and a boom is not available or convenient, a clip-on microphone (see page 50) could be used but is not generally recommended because the speech quality is likely to be variable and inferior, owing to the unnatural position of the microphone.

If the speaker is seated at a table or desk a neat microphone may be acceptable in shot. For good speech quality the top of the table should be covered with acoustic absorbing material (e.g. thick felt) to prevent a booming effect caused by reflection from the surface. A still better arrangement is to make the table out of a framework of wood or metal with a fine mesh surface covered with an open-weave fabric so that the sound waves pass through as though it did not exist.

If the speaker is inclined to thump the table the microphone should be suspended from a floor stand and pass through a hole in the table or behind a masking front piece.

For stereo the microphones could be 'panned' to each side to match the picture, but should not be spread more than about 15°. The level must be kept up for the benefit of viewers with mono sound.

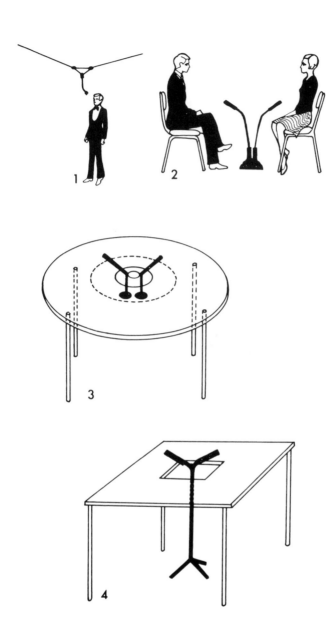

1. Electrostatic microphone with extension tube suspended over static speaker.

2. A pair of electrostatic microphones with short extension tubes mounted on a double stand for two seated speakers.

3. Circular talks table. The table top is acoustically transparent. A sprung platform is provided below the central hole on which to mount the microphone stand.

4. Table with hole for use with floor-stands.

Informal Discussions

When a number of people take part in an informal discussion it is often required to make the arrangement look like a domestic situation. In these circumstances microphones would look incongruous in shot so the situation usually calls for the use of a boom or booms. However if the discussion is completely unscripted the director may wish to have a wide angle shot available on one camera throughout so that he can cut to it at any time for reaction or because he does not know who is speaking. In this case it may be necessary to employ super-directional microphones, on booms kept above the frame of the long-shot, or to hide microphones in shot. A typical arrangement for an informal discussion would be to have the participants seated in a semicircle around a coffee table. This could accommodate a vase of flowers in which could be hidden several microphones pointing towards the various speakers.

The reason for having more than one microphone in these circumstances is to enable a balance to be obtained between voices of different strengths and also possibly at different distances.

With all these arrangements it is likely that the microphones will be working further away from the speakers than is ideal, so it is advisable to have available a 'presence' filter to restore the presences (see page 110).

Excessive microphone distances are also liable to spoil the effect of an intimate setting by introducing a distant, reverberant acoustic effect. As much sound-absorbing material as possible should be incorporated in the settings. It can also be an advantage to place thick drapes on frames to fill the spaces between the cameras. As well as helping the acoustics, these help the participants by promoting a more intimate atmosphere.

If the speakers are likely to move about or change their direction of voice projection radically it may be necessary to fit them with personal (clip-on) microphones as an additional safeguard.

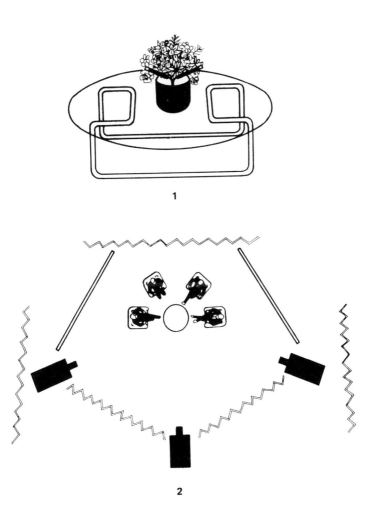

1

2

1. Vase of flowers hiding microphones. Two directional microphones are used instead of one omnidirectional to assist in balancing between voices of different power.

2. Set-up for informal discussion in armchairs. The sets are made of absorbent material and thick velour drapes on frames are used whenever possible to insulate the area from the studio acoustic.

Formal Discussions and Panel Games

Formal discussions and panel games tend to adopt rather similar arrangements in television. In either case the participants and the chairman are usually seated at a table or stylised desk. If it is a game (or discussion) in which the people taking part are competing individually (or expressing individual points of view) they are usually arranged to be seated at one desk with the chairman at one end or at another desk facing them. On the other hand, if it is a team game (or discussion), they tend to be divided, possibly with separate desks with the chairman seated in the middle. In either case it is generally necessary to give each person a separate microphone in order to balance the levels of the voices and to ensure that each microphone is reasonably close to the person concerned.

This is especially necessary if an audience is involved to guard against PA instability (see page 128) and because it may be necessary to compromise for different angles of projection of the voice. The artists may have to turn one way to face the chairman, another to face the audience, and possibly they will also want to turn to talk to each other. If considerable changes of angle are involved it may be necessary to employ extra microphones at the extreme ends of the desk or even, as a last resort, to use personal (clip-on) microphones.

There is usually no objection to showing microphones in shot for this type of programme. Small desk stands can be useful, but there is always a risk of people banging the desk and causing noise by conduction — a risk that is multiplied by the number of persons sharing the desk.

This problem can be overcome by using floor stands arranged so that the microphones enter through holes in the desk, without touching it, or stand in front of the desk. A masking screen hides the stands in longshot views.

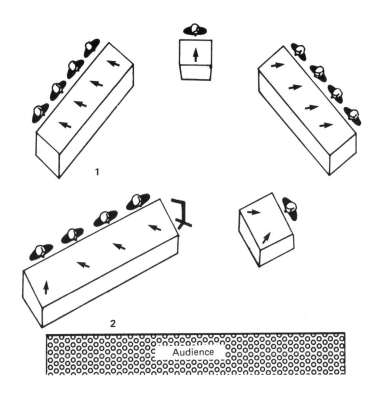

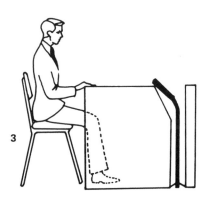

1. Typical grouping for two opposing panels and chairman. Note angling of microphone to take account of artist's projection.

2. Panel game with audience.

Note requirement for extra microphones to take account of artist's alternative direction (to chairman or audience).

3. Section view of panel desk showing microphone stand screen.

125

Planning for Studio Audiences

Many variety and situation comedy productions involve a studio audience. Notwithstanding the fact that most television studio audiences are admitted free they should be treated with the utmost respect and their presence taken into careful consideration at the planning stage. A happy audience is not only a good public relations exercise but also a vital complement to the production.

The audience should be arranged in as compact a formation as possible. This is to assist with PA and audience reaction coverage (see page 132) and because they tend to stimulate each other better in close formation.

Visibility
The action should take place as close to the audience as possible. This is usually quite easy to arrange for variety but difficult for situation comedy if a large number of sets is involved. It is useless to arrange the action all around the studio walls remote from the audience and then expect them to react. Even if the action is reasonably close to the audience great care must be taken not to mask peoples' view more than absolutely necessary. Booms tend to be the worst offenders from the masking point of view and when these are used they should, whenever possible, work from the sides of the set.

Operator discipline
All operators involved in audience shows should avoid distracting the audience's attention when they are not immediately involved, e.g. preparing for the next sequence. The audience can easily find the 'mechanics' of the operation more fascinating than the performance. All 'off stage' movements during a performance must be as unobtrusive as possible and preferably screened from view.

Production routine
The audience should be ushered in to subdued background music (non vocal), and a microphone should be made available to the producer before the show for 'warm up' and afterwards to thank them and send them away happy.

126

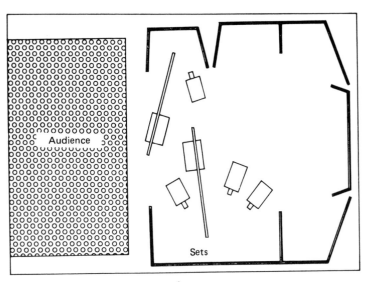

1

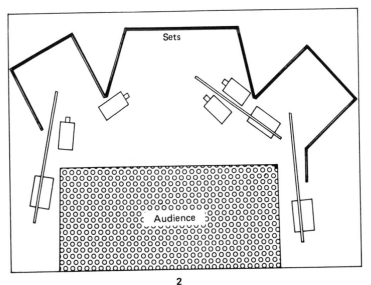

2

1. Wrong arrangement of sets for a situation comedy programme. The action is mainly screened from the audience who get a distracting view of the technicians.

2. Correct arrangement for programme audience. The cameras and booms work from the side as much as possible.

Audience Reinforcement (PA)

One of the most essential aspects of any production involving an audience is an efficient system of audience reinforcement or, as it is usually called, public address. There are two basic types of system.

High level system
For high-level operation a small number of directional loudspeakers (sometimes only one) operate at high level to supply an audience at some distance. The main disadvantage of the high level system is that feeding a high level of sound into the auditorium tends to excite the reverberation and to turn it into a vast echo room. Also, although directional loudspeakers are used, any small reflecting surface in the vicinity can spoil the directivity; at high level this might cause feedback to the microphone, resulting in howlround.

Low level system
For low-level operation, a very large number of small loudspeakers operate at comparatively low level to supply a small section of the audience. These are usually simple small box baffle loudspeakers. Some systems even provide a loudspeaker fitted in the back of each chair. The most common type of directional loudspeaker used for PA systems is the column or line-source. This consists of a number of loudspeakers arranged in line in an acoustic baffle and connected to operate in phase. The back of the baffle box is usually either completely closed or provided with an acoustic resistance. This causes the sound to take approximately the same time to emerge from the back as to travel around from the front so that the two tend to cancel at the back.

The line source loudspeaker tends to have a circular polar diagram at right angles to the width of the column and a narrower lobe along its length.

Line source loudspeakers tend to be more directional in the higher frequencies. This can be counteracted by various means such as curving the column, tapering the input to decrease towards the ends by means of a transformer, or using more closely spaced treble loudspeakers.

Compromise arrangement
In television, where microphone distances can be excessive, the PA problem is often critical and can usually best be met by a compromise arrangement of directional loudspeakers used in large numbers almost in the low level manner. In television studios the best method is usually to employ four-foot column loudspeakers mounted in stirrups that can be conveniently suspended over the auditorium on special hoists or on spare lighting barrels.

128

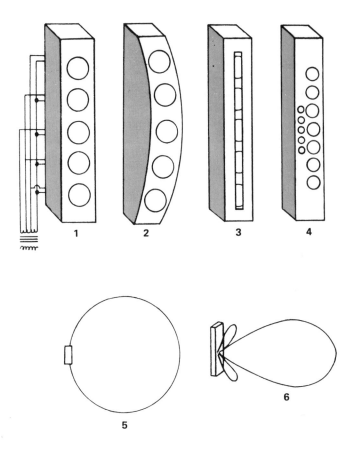

Line source loudspeakers. Methods of balancing the treble and bass directional characteristics.

1. Tapering the input to favour the middle.

2. Curving the face of the box.

3. Causing the loudspeakers to speak through a slit (this widens the HF polar responses).

4. Employing closer-spaced HF units (less directional effect).

5. Polar response of column loudspeaker in horizontal plane.

6. Polar response of column loudspeaker in vertical plane.

PA Connection Points

The need for an efficient system of PA and suitable loudspeaker arrangement was discussed on page 128. There remains the matter of how to connect it to the output of the broadcast microphones.

Before the fader PA

In modern practice programme control is exercised by the individual channel faders. The PA is therefore arranged to be connected at full strength immediately the faders are moved off the backstop.

The main advantage of this system is that the PA is not subject to programme control, so that the loop gain of the potential feedback system (microphone–amplifiers–loudspeakers–microphones) remains fixed and unless the position of the microphone is altered it should be possible to work close to the howlround point without undue risk. Also the fact that the artist is able to hear his volume increase as he approaches the microphone can encourage him not to stray too far from it. The disadvantages are that the sound on the PA may not match the broadcast sound, so any pick-up on the broadcast microphones may be detrimental. Also any sound that is established before the fade-up will come in with a sudden burst.

After the fader PA

If the PA is connected after the fader, it is subject to programme control unless this is exercised only by the group fader. This can present an increased risk of howlround or variable colouration. It can also result in a rather disturbing effect for the audience if the nearest loudspeaker is not in direct line with the artist. The sound can appear to hop about between the two as the relative volume varies.

Combined arrangement

A method that can combine the best features of before and after the fader PA is to divide the control into two sections and feed the PA from the connection between the two. In this way it becomes subject to the lower half of the control function and therefore is faded up gradually but is not affected by subsequent programme control.

The ideal arrangement provides PA connection that can be switched between mid-fader, after the fader, and after the group fader.

Digital PA processors

Digital processors are available which introduce delay and frequency shift into the PA system. *Delay* can be used to correct for the relative distances from the audience of artist and loudspeakers. Delaying the PA (by about 3 ms per metre or 1 ms per foot of the difference) can restore the impression that the sound is coming from the artist and not from the loudspeakers (see page 116). The overall loop frequency response of a PA

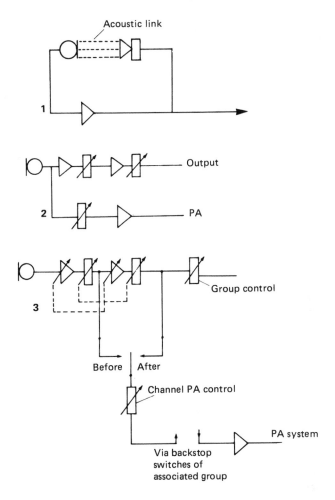

Acoustic link

1

Output

PA

2

Group control

3

Before | After

Channel PA control

PA system

Via backstop
switches of
associated group

1. A public address system is intrinsically a feedback arrangement. The liability to instability depends on the loop gain of the amplifiers overcoming the acoustic path loss. Obviously it is possible to work nearer to the 'howlround' point if both factors are fixed.

2. Simple before-the-fader arrangement.

3. Sophisticated arrangement providing a choice of before or after the fader PA.

system usually has considerable peaks and troughs. *Frequency shifting* the PA output by a few hertz in either direction can effectively move microphone peaks into loudspeaker troughs, thereby reducing the threshold of howlround, which is excited by peaks in the loop gain.

131

Audience Reaction

Audiences are, in general, invited to television shows for two purposes:
1. To encourage the artists and assist them in judging the effectiveness and timing of their material.
2. To provide laughter and applause for the broadcast. The first requirement therefore is that they can see the artist clearly (see page 126) and the second that they can be clearly heard.

Sound character
The type of sound usually aimed at for audience reaction involves two characteristics that tend to be conflicting:
1. The audience should sound large and impersonal (i.e. no individual should be prominent — especially if he has a distinctive laugh).
2. The audience should sound reasonably close and intimate in character, ideally giving the listener the impression of being seated in the middle of the audience. This effect can usually only be obtained using a large number of microphones, each one at a distance of only 2–3 mm (6–10 ft) from the particular group of audience it is intended to serve.

Separation from PA
A major problem in designing an audience reaction microphone set-up is minimising pickup from the PA loudspeakers which are serving the same area. Sound picked up in this way adds colouration from the uneven response of the PA loudspeaker to the broadcast sound and makes it seem reverberant, turning the auditorium into a sort of large echo room. This can best be avoided by using directional loudspeakers and microphones with their dead sides facing each other.

'Milking' the audience
It is unlikely that the desired effect of audience reaction will be obtained for most comedy production. Applause is usually much louder than laughter but should be broadcast at a rather lower volume. Some weak laughs need to be magnified somewhat.

It is therefore often necessary to 'milk' the audience. That is to 'swell' the volume of laughs by bringing up the volume after the laugh has started, increasing the natural rise and fall, as realistically as possible. It is important to reduce the level before the artist speaks again otherwise his voice will be picked up by the reaction microphones, producing an echo effect.

Compressors (see page 104) can provide a useful way of increasing the average level of reaction without risking overload, but means must be provided to cut back their output for applause, otherwise it will be much too loud.

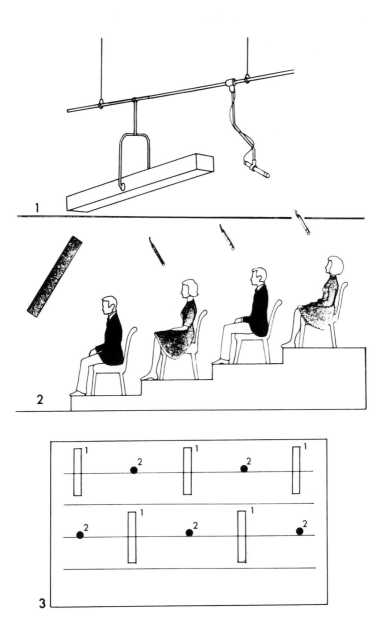

1. Arrangement of PA loudspeaker and audience reaction microphone on lighting barrel.

2. Microphone arrangement for a more distant loudspeaker (high-level) system.

3. Plan view of audience area showing disposition of loudspeakers and microphones.

Audience Participation

Some of the most difficult types of discussion programmes from the sound point of view are those where members of the audience are invited to take part. It is quite easy to arrange for a camera to zoom in from a wide-angle shot of a large audience to show a close-up of a single speaker, but picking up the corresponding sound can be a very different matter.

Set-up arrangement
The simplest set-up occurs when it is possible to arrange that only a few specified members are allowed to speak. These can then be placed in convenient positions where they can be covered by suitably suspended or stand microphones.

Hand-held microphones
In circumstances where any member is able to speak, but there is good notice of the order, it may be possible to pass around a hand held microphone. A number should be made available, each to cope with a small block of seating.

Shotgun microphones
If the discussion is *ad lib* but not very quick-fire, coverage can often be achieved by planting sound operators with hand held shotgun microphones among the audience, each to cover a few rows of seats.

Booms
The best arrangement for a free-flowing discussion is usually to use booms equipped with shotgun microphones. The reason for this type of microphone is that the booms are unlikely to be able to reach all members of the audience and will probably have to work at a height sufficient to allow the cameras to be able to take wide-angle shots without seeing too much of the booms or their shadows. Several booms will be needed to cover a large area. This also enables one boom to search for the next likely participant while another is covering the current speaker.

Parabolic microphones
Where the microphone coverage has to be inconspicuous, and provided that the auditorium is fairly short and steeply raked, good results have been obtained using a combination of microphones with parabolic reflectors and slung conventional microphones. The slung microphones provide a basic 'blanket' coverage (as for audience reaction), albeit too far from any individual, and the 'presence' is added by mixing in a suitable proportion of a well aimed parabola microphone from which the bass has been rolled off. In this way it is possible to make the sound appear to zoom in with the camera.

Alternatively 2 m (6 ft) shotgun microphones could be employed in place of the parabolas, carefully aimed at the participant.

134

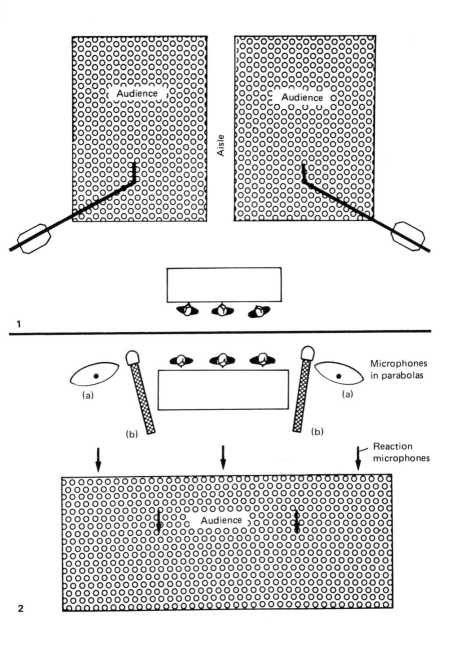

1. Audience grouped into two compact areas each covered by a boom equipped with a 'rifle' microphone.

2. Audience coverage by combination of (a) reaction microphone and parabolic reflectors or (b) 2 m shotgun microphones.

Discussions between Remote Centres

In television it is quite often required to link up people in different areas in such a way that each can see, or at least hear, the other and the viewer can see and hear both.

Deaf aids
The major problem in achieving two-way communication using loud-speakers at each end is to prevent so much pickup of the loudspeakers on the microphones that the sound will have a hollow coloured quality. With loudspeakers at both ends it is possible even for the whole loop to go into self oscillation with disastrous results. The simplest way of overcoming this is to provide the participants with deaf-aid earpieces (less obtrusive than earphones). Earphones or deaf aids are essential when long-distance link-ups are involved, because of the time-lag in the transmission.

Loudspeakers
If loudspeakers are used the following rules should be adopted:
1. The loudspeakers should be fed from a 'clean feed' source. This supply, as its American term 'mixed-minus' more graphically suggests, provides each participant (and there could be a number of sources) with everybody's sound output except his own.
2. Directional microphones, and if possible directional loudspeakers, should be provided, each with its directivity pattern exploited to obtain maximum separation from the other. In practice it has been found that in reasonably dead acoustics the best results are achieved by using a cardioid microphone as close to the artist as possible, with a loudspeaker at a distance of at least two metres (six feet) pointing toward its dead side.
3. The settings surrounding the artist must be as sound absorbing as possible and not so angled that they reflect the loudspeaker sound back on the live face of the microphone.

Line-up routine
When several sources are involved, each using loudspeakers, it is essential to adopt a strict routine in which the loudspeaker levels are each set for just comfortable listening before their respective microphones are faded up. Any subsequent re-adjustment could cause one location to 'steal' more than its share of the available volume and cause the others to reduce their gain to avoid a howlround.

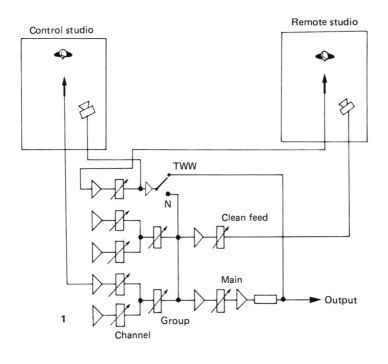

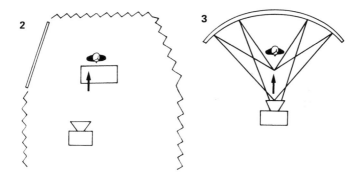

1. Arrangement for deriving 'clean feed' (American term 'mixed-minus') for two-way working with a remote source.

2. Sensible arrangement for remote interview studio. The loudspeaker is at some distance from the dead side of the microphones. The settings are as absorbent as possible.

3. Bad arrangement: the concave reflective settings focus the loudspeaker output on the live face of the microphone.

Planning for Drama

Planning the sound coverage of a drama production should begin many weeks in advance of the studio date by a careful reading of the script and discussion with the director to find out:

1. The character of the story, the period, pace and general degree of sophistication and complexity to be employed in the production.
2. The requirement for filming or electronic pre-recording, possibly of exterior sequences, and the need for matching with the studio sequences.
3. The requirement for incidental music and recorded sound effects, and whether they are already available or will need to be specially recorded in advance.
4. The need for special effects or 'props' such as 'practical' (working) telephones or intercoms etc.

Filming or pre-recording

If filmed or pre-recorded sequences are involved and required to match the studio sound it is advisable to find out which medium is likely to be faced with the least flexible situation and then arrange for the other to adopt a similar technique. For example it is no use the studio taking advantage of the opportunity to use very close microphone technique if the pre-recorded sequences with which it is to be intercut are forced to work at a distance. A compromise should be adopted to make the transition less obvious.

Planning meeting

The sound supervisor should attend at least one full-scale planning meeting several weeks in advance of the production. This would include the director, the lighting director, scenic designer, wardrobe and make-up.

The sound supervisor should check the arrangements of the sets and the camera positions and, in consultation with the lighting director who can give an indication of the direction of the key lights, decide on the best method of obtaining sound coverage.

This is achieved by reference to the plans, elevations and models as outlined on page 92. The space occupied by the boom and its coverage can be ascertained by the use of a Perspex cut-out. It may be necessary to arrange the juxtaposition of the various sets to assist in boom movement or to provide sufficient acoustic separation for perspective effects etc. This is usually achieved by cutting out the outline of the sets on tracing paper and moving the pieces around on the plan. When the position of booms and other microphones is decided they are marked on the camera plan.

Once the sound coverage has been established the sound supervisor can mark up his script to show the sequence of sources.

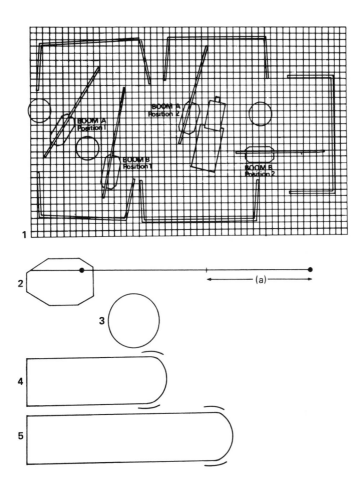

1. Studio floor plan, to the scale of 6 mm to 30 cm (approx. ¼ in to 1 foot), used with stencil cutouts of production equipment for planning.

2. Large boom: (a) range of extension 3–6 metres (10–20 ft).

3. Vinten camera pedestal.

4. Vinten Heron camera crane.

5. MPRC camera crane.

Dialogue and Split Sound

Most drama involves dialogue, i.e. people talking to each other. If the coverage is by means of a boom it is usually possible to cover a number of people with one microphone by splitting the difference between them, provided that:

1. They are not much more than about two metres (six feet) apart.
2. The shots of the individual actors are not extremely close and the dialogue intimate.
3. The people are reasonably equidistant from the cameras and of fairly similar height.

When these conditions are met the usual position for the boom is above and downstage, i.e. toward the cameras but splitting the angle between them, unless it is necessary to favour one of the artists to compensate for a weak voice or match an exceptionally close shot.

Split sound

Problems can arise when the artists are separated by distances much greater than two metres. If they merely go towards opposite sides of the set and are pictured entirely in long shot it still might be possible to cover both with a boom held high between the two. However, as soon as a camera takes a close-up shot of one of them it will be necessary to bring the boom in close to cover him. If another close-up shot follows quickly it is most unlikely that the boom will be able to swing over in time. It will therefore be necessary to 'split the sound' by bringing in another boom so that each can cover a section of the area in turn.

Upstage–downstage split

While it is usually quite simple to use two booms on one set with a sideways split, it may not be so easy when the artists split between front and back of a deep set so that one is seen over the shoulder of the other. In this case there is considerable risk of seeing the upstage boom in the shot of the downstage artist. It can be useful to have available another method of picking up the sound close to the upstage artist, e.g. a personal or hidden microphone which can be mixed in with the output of his (too high) boom to restore presence in the close-up.

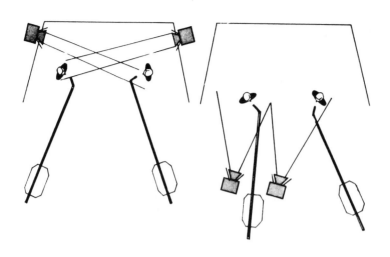

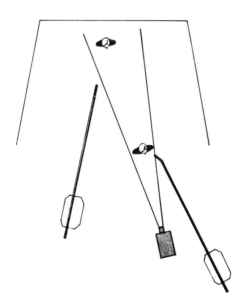

1. Boom positions for split sound with three-quarter back key lighting.

2. Boom position for split sound with frontal keys.

3. Upstage, downstage split. Left hand boom is too high for close-ups and is augmented with output of personal or hidden microphone.

Grouping Problems

When crowds or large groups are involved in drama productions they are usually composed of extras with no lines to say except the proverbial 'rhubarb-rhubarb' background chatter. This can produce a very unrealistic effect unless they are told to speak natural conversation at a natural level of voice. The difficulty then usually arises that the actors, who may be close to the crowd, or even mixed up with it, have to be clearly heard above the crowd; moreover, no irrelevant dialogue from members of the crowd must be sufficiently distinct to be distracting. The best way to achieve a natural effect in these circumstances is to make a recording of the crowd using natural voices but with such an overall balance that the words tend to merge together. This recording is played back to the crowd during the action and they are instructed to go through the motions of speaking with very little sound so that their live voices merely augment the recording. The sound supervisor then has complete control and can arrange for the actors' dialogue to overlay the realistic crowd noise with sufficient clarity. It is usually advisable to employ two booms, one to remain fairly high for an overall atmosphere effect (the extras movement noises may require synchronous sound more than their blended voices) and another to drop in close to the actors when in close-up.

Tables

Quite often the action in television drama centres around a table. It is usual to arrange the actors, or at least those who speak, around three sides to allow access for the cameras. This usually results in a deep but close form of grouping in which it is very difficult to obtain close-up sound to match close-up pictures. The problem is to avoid shadows (from the close cross-key lighting) and the necessity for the boom to rack and turn quickly. It is usually advisable to employ two booms, each looking after a defined area. If long shots are involved it may be possible to hide a slung microphone in a lamp fitting over the upstage end of the table.

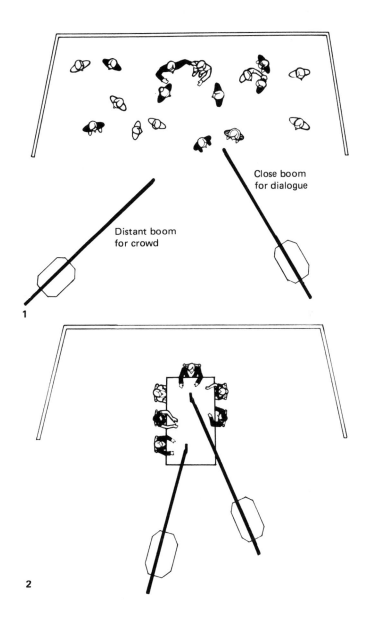

Close boom
for dialogue

Distant boom
for crowd

1

2

1. Crowd scene. Two booms are used, one close-in for dialogue and one high for general crowd atmosphere noises.

2. A deep arrangement around a table. Two booms may be required but the 'upstage' one has to work at some height to avoid throwing shadows on the downstage actors.

Problems of differing heights between artists can often be overcome by working the microphone to the side of the frame.

Height Discrepancy

A problem can arise when there is dialogue between somebody standing up and somebody seated, especially if the standing person has the stronger voice. If the shots are not too close it may be possible for the boom to work from a position well to the front to take advantage of the cone angle of the camera lens, or to work to the side of the frame nearest the seated person.

Long shot to upper room or balcony

There are occasions when dialogue occurs between someone at ground level and a person upstairs, perhaps on a balcony or staircase, or in an upstairs room. Often it is required to see the higher person over the shoulder of the lower one. This is likely to be a split sound situation, the upstage actor taken on a boom tilted high (the maximum height of a boom microphone with the arm at full rack is about 5 metres, 16 feet) or with the microphone hidden in some convenient set dressing such as a lamp fitting or balustrade. It may be difficult for the other boom to approach the downstage actor sufficiently without coming into shot, in which case it might be advisable to work to the side of the frame or use a fishpole or shotgun microphone held underneath the shot.

Action on rostra

When the action takes place above studio floor level (e.g. when low angle shots are called for, or the action has to be continuous to an upstairs location) it may be necessary for some of the cameras to work on the rostra also. If there is not room for the boom to do so as well it may be possible to fix to the scaffolding supporting the camera platform a spigot on which the boom arm can be mounted. This is rather a restricting situation and can be an unnatural operating position, so on most occasions a fishpole is preferable.

144

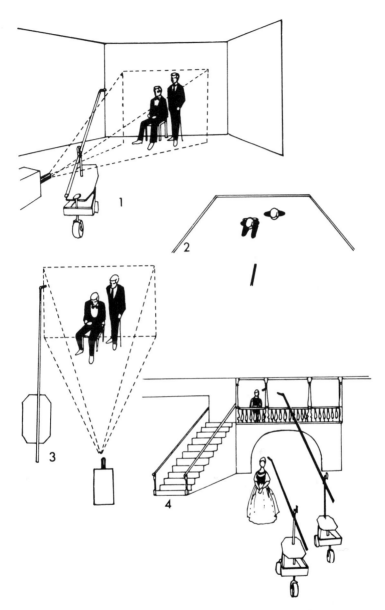

1. Boom operating well forward to take advantage of low-angle shot.

2. Plan view showing approximate position of microphone.

3. Boom working to side of frame.

4. Close-up and long-shot coverage for high split-sound situation. A microphone behind the balustrade can be added for 'presence' wben the upstairs actor is is in close-up.

Ceilings, Beams, Windows

It is sometimes desirable to provide sets with a ceiling, or at least a partial ceiling, to obtain a more realistic long shot or allow for a low angle upward-looking shot. This can present a considerable problem for lighting, owing to the difficulty of back-lighting the artists (see page 86), and for sound. One way of overcoming it is to make the ceiling from a succession of coffers or beams which, when viewed from the front, look like a continuous ceiling but which allow space between to shine backlight and possibly to position fixed microphones to cover the action underneath. Alternatively a thin gauze can be stretched over the area. If it is light in colour and lit from the front it will appear to be solid, but a boom or other microphone can be used behind it very effectively. It is, however, difficult to shine backlight through a gauze without spoiling the effect and exposing the position of the lamp. Where low beams are involved in the set it may be possible for a boom to work effectively over the top of them, where possible working parallel, between the gaps.

Windows

Windows in television productions usually consist of empty frames without glass. Obviously the frame has no effect on the sound picked up by the microphone, but if actors are to be seen speaking their lines apparently through a closed window it is necessary to preserve the illusion of glass by modifying the sound. This can be achieved most effectively by dropping the sound level to about half the normal and providing high frequency attenuation of about 3 dB per octave whenever the sound is supposed to be heard through the glass.

Reverse angles

Quite often windows at the back of the set are used to provide a reverse angle viewpoint for the camera. In this case it is necessary to provide a microphone outside the window in line with the camera shot, and fade to it when that camera is used, to prevent a reversal of perspective between picture and sound.

Stereophony

Where stereophonic sound is involved there is also the problem of left-right inversion with reverse-angle shots. In this case the lateral spread should be severely limited to prevent the sound becoming 'jerky' if it follows the vision.

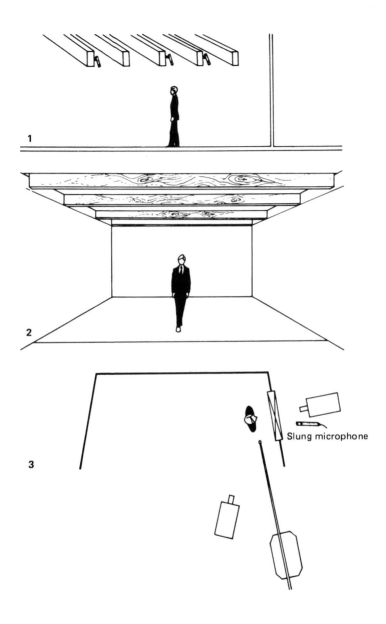

Slung microphone

1. Side view of set showing suspended beams, some of which hide microphones.

2. From the front the beams take on solid appearance.

3. Reverse angle shot through window showing the need for a microphone in line with the camera.

Columns, Arches and Doorways

Columns are sometimes included in television design for architectural reasons or merely to give a sense of depth to the set. They may be supporting scenery or standing on their own, possibly steadied by wire from above. They can cause considerable problems for the boom, particularly if the artists walk past them while speaking. If the artist walks close to the columns, whenever possible the boom should work from the same side as the artist. Even then there is a considerable risk of microphone shadows being cast on the columns, as the lighting will probably also be from that side, so that the boom may have to work at some distance in front of the artist. If the boom has to work from the side of the column it will be very difficult to prevent a break in sound continuity as the boom has to rack in, pass each column in turn, and rack out again. In these circumstances it should either be arranged that the artist does not speak as he passes the column or two booms must be used, one taking over from the other.

Arches
A similar situation occurs when an artist has to walk though an arch while speaking. If the shot remains of constant length from the front so that the artist approaches the camera it may be possible to cope with one boom on the camera side of the arch, but if the camera tracks through the arch, or zooms in to close-up on the other side, two booms will be necessary to cover the walk, one either side of the arch. In this type of situation it is a great help if the sound supervisor can see into the studio so that he can judge the best point to make the change-over. If he is not in a position to see into the studio the boom operator should warn him over the reverse talkback as soon as he has handed over to the other boom and can be released.

Doorways
Doorways present a similar situation to arches except that there is more often the possibility of a reverse angle shot from the other side of the door. This is usually a convenient place for a fixed-position slung microphone.

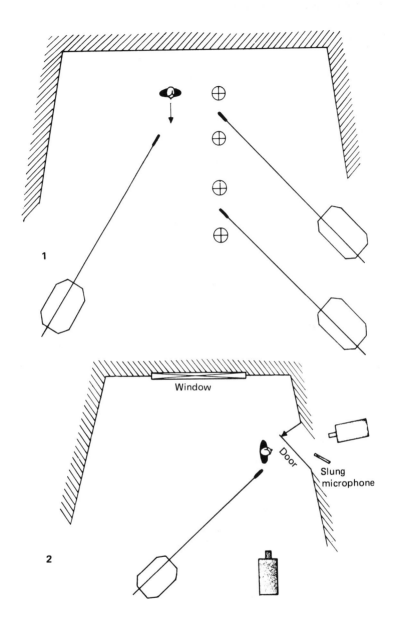

Window

Door

Slung
microphone

1

2

1. If possible the boom should work on the same side of the pillars as the artist. If this is not possible, for lighting or other reasons, two booms may be necessary.

2. When an artist works at a door, or shots are taken through it, it is usually necessary to employ another microphone outside the set. This could be a slung microphone.

Vehicles

Television drama often takes place within the confines of vehicles.

Railway carriages

Railway carriages are usually small cubicles with parallel sides. To overcome the problem of taking shots of the actors on both sides of the carriage it is common to have the cameras looking through gauze pictures or trap doors above the opposite people's heads on either side. This and the frequent requirement to take overall longshots to include the window, with moving back projection or colour separation overlay (see page 152) to give the illusion of movement, tend to make it difficult to keep the microphone out of shot. Nevertheless a boom is usually the best answer, possibly with a super-directional microphone (see page 52) to make up for the excessive working distance. The resulting rather 'boxy' acoustic tends to add realism. The sound effects, which are supplied by tape or disc, help to cover any increase in noise level. These have to be carefully synchronised to match the apparent speed of the train, and increased in volume if the compartment door is opened.

Lifts (elevators)

Lifts present little problem when shot in the usual way from eye level as though from the door. There can, however, be a temptation to shoot them from above via a mirror to give a claustrophobic effect. In these circumstances it may be necessary to resort to personal microphones on each of the artists. They should be equipped with radio transmitters if they have to move, but care must be taken that they do not interact with each other in such proximity.

Motor vehicles

There is a frequent requirement to shoot action in motor vehicles of various types, particularly police cars. Usually the cars are seen against back projection (or a CSO screen, see page 152) visible through the side or rear windows, sometimes both. Obviously the sound effects have to synchronise exactly with the apparent movement of the car and the gear changes etc. as well as relate to the other traffic that can be seen. There is a considerable advantage if the film or pre-recording used for the background can be shot with synchronous sound, otherwise great skill is required of the gramophone operator. In either case the sound effects should be fed to the actors on a loudspeaker to assist them to synchronise their gear changes etc.

Picking up dialogue from the front seats is fairly simple with a boom if the windscreen has been removed. Otherwise a useful expedient is a small electrostatic microphone mounted on a strip of flat springy material such as laminated plastic which can be sprung into position between the sides to fit flush with the roof. Alternatively, an acoustical boundary microphone can be attached to the roof.

150

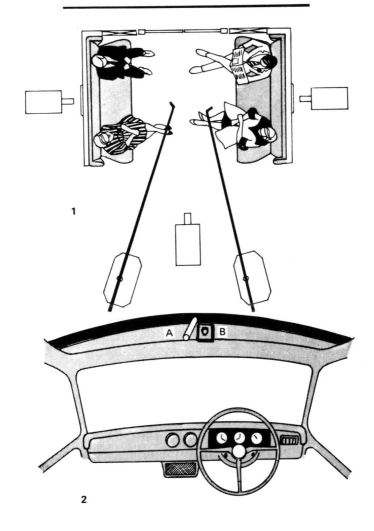

1. Typical setup for railway carriage effect. The two booms have to work fairly high to clear camera shot through gauze windows above artists' heads.

2. A laminated plastic strip is sprung into the roof lining of a car to suspend a microphone, A. A loudspeaker may be necessary if 'intercom' is required with a remote 'base station'. B: an acoustical boundary microphone attached to the headlining.

151

Back Projection and CSO (Chroma Key)

Some senes, notably those which include an exterior background, are either too complicated to fabricate in the studio or are required to give the effect of movement, in which case photographic or electrically generated backgrounds are used.

Back projection

When back projection is required the action takes place in front of a large translucent screen through which a photographic image is projected from a still slide or motion picture film projector.

The sound problems in back projection usually stem from the noise of the film projector. Obviously it helps if the projector is properly 'blimped', but even so whenever possible the whole apparatus should be built into an acoustically treated box.

In designing the set-up for back projection the projector should be arranged to be as far as possible from the acting area and, if possible, heavy velour curtains should be draped to form a screen between them. Some BP projectors are equipped with an outlet of sound (usually optical). This is probably not of very good quality, but may be good enough to provide sound effects synced to the film. Alternatively the optical sound track can be used to help in synchronising a separate magnetic track, run in telecine.

Colour separation overlay (chroma key)

In modern practice back projection has been largely superseded by an electronic process called colour separation overlay (CSO). In this system the area into which the background is to be inserted is brightly lit and painted in a pure primary colour, usually blue or yellow.

An electronic switch is incorporated in the output of the camera providing the foreground, and matters are arranged so that wherever this camera sees this colour its output is automatically exchanged for another source such as film (from telecine) or the output of another camera, either live or prerecorded.

In this way the artists, and any foreground objects not of the CSO colour, are inserted into the background (i.e. unlike a simple superimposition, the background cannot be seen 'through' the foreground).

CSO does not normally present problems for sound, and in some circumstances could be used as a positive asset. For instance, if sound is required from the 'background' source it may be possible to find a place for the boom which, though well in shot on the background camera, is not seen in the final inlaid picture.

Care must be taken to match the acoustic and environmental effects of the foreground and background sound to the composite picture.

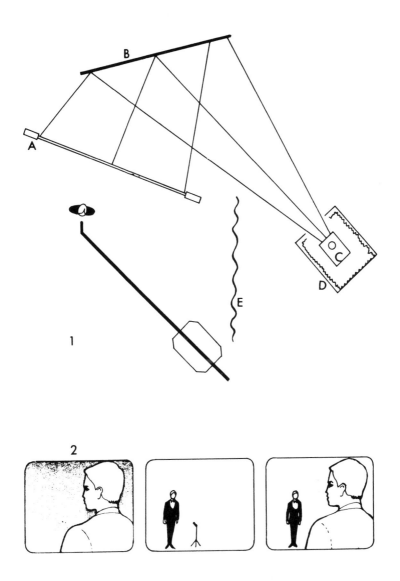

1. *Back projection.* Typical setup for a back projection sequence. A, BP screen; B, mirror (folding-back the beam to accommodate the length of throw); C, projector; D, sound absorbing box; E, heavy velour curtains.

2. *Colour separation overlay.* The foreground picture (left) is shot against a background of the CSO switching colour. The background picture (centre) is composed to fit the switching colour area. Note stand microphone in shot on the camera. In the combined picture (right) the foreground has overlaid the background and the microphone is now hidden.

Weather and Sound Problems

Wind

Most sequences shot out of doors are subjected to wind. Pseudo-exteriors shot in the studio can be made to look much more realistic if the foliage and the artists' hair and clothing are blowing about. Unfortunately wind on the microphone does not sound like wind at all. Usually it is more like thunder.

The first essential, therefore, where wind is likely to be encountered is to have effective windshields on all microphones. For this purpose the small close-talking shields, usually made of plastic, supplied with many microphones are of little use. To be effective a windshield needs to be at least 10 cm (4 in) in diameter with a smooth, round or streamlined shape. It must be mounted firmly to the microphone within its suspension so that the microphone cannot move inside it.

Wind machines

When wind machines are used, care must be taken in planning for the type of machine and its location to localise the wind and keep it off the microphone.

The so-called silent wind machines which use a multi-bladed variable-speed fan mounted on a wheeled trolley can, with care, be used for sound sequences. When only a close-up shot is required, sufficient wind over a small area can be obtained from a small domestic fan, mounted on a stand or hand held.

Trick lines

Where large movement of foliage is required it is always worth considering whether a more realistic effect could be obtained by using a 'trick line' to shake the branches. In any case the effect should be augmented, and any machine noise covered, by a recorded effect of wind noise.

Rain

Where 'practical' rain is involved there are two main problems for sound. One is to prevent the microphone getting wet. This can be achieved by enclosing the microphone in a thin plastic bag within a windshield. This is to prevent the sound of the rain hitting the plastic bag. Surrounding the microphone with a continuous thin membrane in this manner tends to destroy its directional characteristics, and can result in turning a cardioid microphone into an omnidirectional one, but should not otherwise affect its response.

The other problem associated with rain is noise, particularly the nosie of water dripping from roofs and sills. It is advisable to use sacking or tightly rolled chicken wire to break the fall of the water, especially if it is supposed to be on the other side of the window.

154

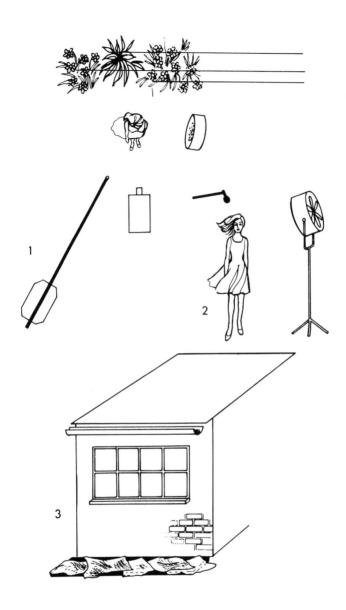

1. A windy 'exterior' wcene. Note the 'trick lines' to shake the foliage and the small wind machine to rustle artists' hair and clothing.

2. Front view shows how the fan should be tipped down to prevent wind blowing on to microphones.

3. Roof and window prepared for a rain sequence with tightly rolled chicken wire in guttering and sacking on drip tray on ground.

Gunshots have to be made to sound loud by acoustic quality rather than volume, which is limited by the transmission medium.

Gunshots

Quite a large proportion of televised plays involve the use of firearms which have to be fired with realistic effect.

Real guns

If the gun has to be seen to fire, the best arrangement is obviously to use a real gun with a blank cartridge so that the flash can be seen to synchronise with the bang. This creates a big problem for sound because in many cases the gun is fired in quick succession to the dialogue and is picked up on the same microphone. When the volume has been controlled to keep within the transmission limits, either manually or by a limiter, it will come out only slightly louder than the speech, a puny shadow of the original. The best way round this problem is to use a limiter in the output of the dialogue microphone (usually a boom) to 'cut the top off' the peak of volume. The limiter must have a short recovery time (see page 104) so that it does not hold back the succeeding dialogue. It does not matter much if the resultant sound is a bit distorted; this can increase the subjective effect of loudness, as the ear tends to distort the sound when close to a loud bang.

At the same time another microphone placed at some distance, possibly high up in the studio, should be faded up (with edho added if the scene is supposed to be indoors), to collect the reverberation and give the sound a 'tail' and thereby an impression of loudness.

Don't be afraid to use a full charge cartridge — the louder the sound in the studio the better.

Electronic effects

Where it is not necessary to see the flash of the gun the effect can be laid in with a tape or disc recording or by means of an electronic gunshot generator. This is an electronic device using 'triggered white noise' (a signal made up of equal proportions of the entire frequency spectrum in the form of an impulse). This can be set to produce an effective representation of all types of small arms effect, and even the characteristic whine of the ricochet, by merely selecting the required effect on a rotary switch and pressing the button on cue.

Foldback

Whenever any artificial method of reproducing gunshots is used it is essential that the artist hears the bang as loudly as possible to induce the necessary reaction.

The sound must therefore be 'folded back' on a powerful loudspeaker situated close to the action.

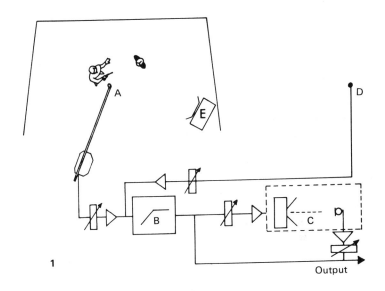

1

Output

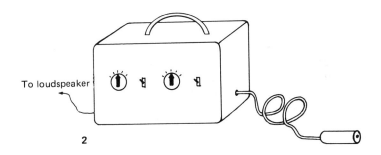

To loudspeaker

2

1. *Set-up for murder!.* A, dialogue microphone. B, limiter. C, echo chamber. D. distant microphone slung high in the studio. E, powerful loudspeaker for use with gunshot generator.

2. General appearance of gunshot generator.

It is an established convention that the distant end of a telephone conversation should sound distorted and at lower volume.

Telephone and Intercom Effects

Telephones are used frequently in drama productions. At the first suggestion that one is to be used the first thing to ascertain is the period and type of telephone required. If it is an unusual variety, e.g. foreign or very old fashioned, it may be necessary to get hold of it well in advance to make it 'practical'. This is because the two ends of the conversation will be in different sets which may be so separated that the artists need the telephone to hear each other.

Telephone distort effects
There is a convention in television drama as in radio that if it is necessary to hear both ends of the conversation the distant end is distorted and lower in volume than the speaker seen in shot. The distortion usually takes the form of a low and high frequency roll-off at about 12 dB per octave from below about 500 Hz and above about 1.5 kHz, depending on the type of voice. The proper volume for the distant end really depends on the length of the shot. In reality only the user hears the telephone but it does not seem too incongruous if the shot is reasonably close. It seems ridiculous to hear it in long shot.

Effects switching
When, as is often the case, the shots are intercut between the two ends of the conversation it is vital that the sound cutting always follows the vision exactly, even if the vision switches at the wrong time. If you see a person apparently in close-up with 'telephone' sound it is natural to believe that the sound has gone wrong. To ensure that the picture and sound exactly synchronise, a relay-operated effects unit can be made to switch the output of any of four sources either direct or through a variable filter and attenuator to create the desired effect. The switching in and out of the effects can be done manually or tied to the vision switcher by means of a series of keys which select a particular camera to switch a particular source. This can result in split-second synchronising so that it does not matter if the vision switcher cuts in the middle of a word. The appropriate sound will follow provided that the sound supervisor has selected the right camera/microphone combination.

Intercoms
When office-type intercoms are required the most realistic effect can be obtained by making the units 'practical' and picking them up on the dialogue microphones.

158

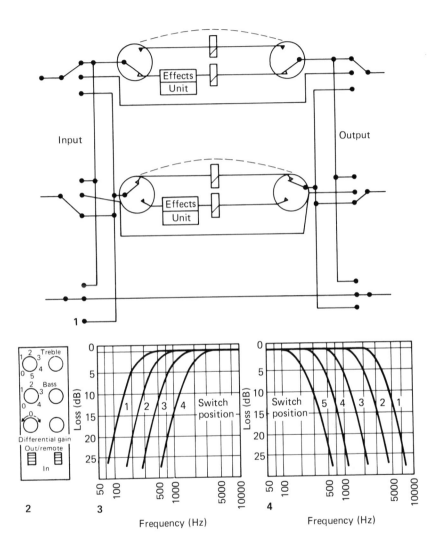

1. *Switching arrangements.* The switches are relay operated and can be worked either manually or in association with the camera-switching relays. Two effects units are used instead of switching one between the desired circuits so that the levels can be balanced and the optimum effect obtained.

2. *Distort unit.* A typical telephone effect distort unit, allowing a choice of different degrees of bass attenuation (3) and high frequency attenuation (4).

159

Exterior Sequences in the Studio

Many dramatic sequences, and sometimes whole dramas, take place apparently out of doors. These exterior sequences often have to be shot in the studio in simulated exterior scenery.

Exterior scenes in the studio are usually shot against a background of a cyclorama lit to resemble sky, or a painted backcloth. In either case this usually consists of a canvas material stretched between poles or runners at the top and on the floor. The main problem is that the outside situation represented calls for an open-air acoustic which in real life would be completely dead except for a few discrete reflections from buildings or trees etc. To make matters worse, a large exterior usually calls for large, wide-angle shots which means that the booms covering the action have to work further from the artists than normal. As there is usually very little material in the vicinity to absorb or break up the sound, the effect will be reverberant — not at all what is wanted for realism. Matters are considerably improved if thick velour drapes can be hung behind the cyclorama cloth. They absorb sound and also help to stop the cloth billowing in the breeze if a wind machine is used.

The artists should work downstage as much as possible to give an impression of space without requiring too much headroom for the boom.

Use of super-directional microphones

It can be a good idea to use shotgun microphones (see page 54) in the booms for these sequences. By including bass roll-off in their output they can be made more directional, thereby reducing reverberation and making the sound more realistic, as sound tends to lack bass at any distance in the open air.

Footsteps

The other problem is the unrealistic sound of footsteps, which can sound just as they are — walking on a studio floor.

If the ground is supposed to be soft it should be covered with a grass mat, real turves, or a thick coating of peat.

If the scene calls for a paved area, and particularly if it is on a rostrum where footfalls would sound 'wooden', paving stones should be laid. Real paving stones are heavy and easily broken, so imitation ones have been designed, made of resin and glassfibre on a metal frame. They are light and easy to handle and can look quite realistic.

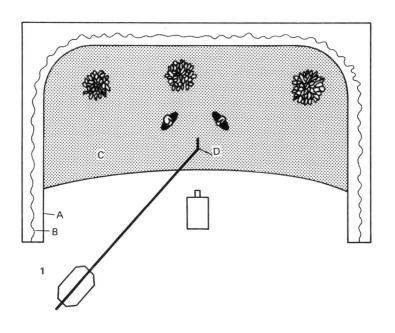

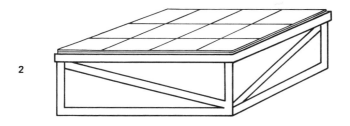

1. A cyclorama cloth. B, thick velours behind cyc cloth. C, turves or peat on floor. D, shotgun microphone on boom.

2. Paving stones (or glass fibre substitute) laid on a rostrum for realistic sound effect.

Location Shooting on Film

In spite of the brilliant efforts of scenic designers, few exterior sequences look quite so convincing when shot in the studio as they can when shot in a suitable exterior location. Film is obviously a suitable medium for much outside location work. This may involve producing synchronous sound track, some of which may have to match in with studio sequences.

Sixteen millimetre film is an excellent medium for location shooting because both the camera and the sound recording equipment (usually a 1/4-in tape recorder such as the Nagra) are small and light and can be manoeuvred into confined locations.

Synchronising methods

The tape recorder and camera can be synchronised with each other by one of a number of systems. Normally a pulse is recorded on the 1/4-in tape either on a separate track from the programme sound or in some cases across the sound track in such a way that it can be recovered separately. These pulses act rather like the sprocket holes on the picture film. In other words they locate a particular element of the sound track with a particular picture regardless of any slipping or stretching that may have occurred to the tape.

Later, in the dubbing theatre, the separate magnetic sound recorder is driven from the pulses recovered from the 1/4-in tape so that it is recorded in synchronism with the original picture film.

The pulses can be fed from cameras to tape recorder by means of a wire connection, but if they have to be separated by a considerable distance a miniature transmitter and receiver can take the place of the wire.

An even more convenient synchronising method is to drive the camera and tape recorder pulses from two identical crystal-controlled oscillators so that they remain synchronous even though completely separated.

The microphone techniques for film recording are very similar to those used with electronic picture production. In practice, the most frequently used device is the shotgun microphone (which is usually battery operated) used with pistol grip and windshield. This makes a simple mobile arrangement that suits many circumstances and matches the 16 mm film camera in manoeuvrability.

This is another reason for using shotgun microphones for exterior studio sequences (see previous section) if they have to match to film.

Film recordists should always record some 'wild track', i.e. periods of nonsynchronous 'atmosphere' and any natural noises that relate to the scene. These can be used to give realism in the editing and dubbing process and can be used by the television studio to promote matching with 'live' material.

162

1

System	Speed (in/s)	Sync pulse frequency	Track positions Audio | Pulse	Head displacement and slit orientation	Track width (inches) Audio	Pulse
Pilote (Maihak)	$7\frac{1}{2}$	50 Hz		Erase / Audio rec / Pulse rec / Audio rep	0.2	0.01
Neopilotton (Nagra)	$7\frac{1}{2}$	50 Hz Europe 60 Hz USA		Erase / Audio rec / Pulse rec / Audio rep	0.2	0.03 total
Perfectone	$7\frac{1}{2}$	100 Hz		Erase / Pulse rec / Audio rec / Audio rep	0.2	0.01 each
Ampex (Fairchild)	15	60 Hz		Erase / Pulse/Audio rec / Audio rep	0.2	0.2
Rangertone	$7\frac{1}{2}$	60 Hz		Erase / Audio rec / Pulse rec / Audio rep	0.2	0.03

2

1. A film location unit at minimum consists of a camera and microphone. The sound and synchronising track from the camera are recorded simultaneously on a small battery tape recorder.

2. Some sound/camera synchronising systems. Time code can be used to synchronise film camera and recorder. For very complex recording it is possible to record time code on the camera in synchronism with a number of tape recorders, each controlled by digital time code.

Shotgun microphones mounted on booms, hand held, or mounted on cameras can be used for electronic location production.

Electronic Cameras on Location

The technique of sound pickup for outside broadcast location shooting is very similar to studio work except in three important respects, weather, noise and difficult circumstances.

Weather

The main weather hazard on outside broadcasts is wind on the microphone. This means that in practically every external sequence every microphone has to be fitted with a really effective windscreen (see page 62). Similarly it may be necessary to protect the microphones from rain.

Noise

Very few outside broadcasts are totally free from noise. Often the type of noise will be quite incongruous to the production, such as aircraft or traffic noises. The noises may be based on a particular frequency, in which case a 'notch' filter or octave band filter can be very useful in recording a period drama. The only real solution is to choose the site and the moment carefully and adopt as close a microphone technique as possible. Then suitable recorded atmosphere and sound effect noise can be added to obscure any incongruous background that cannot be eliminated.

Use of booms

There are some situations in outside broadcasts where it is feasible to use booms, and this is certainly ideal for drama where it is practical. For outdoor sequences the booms are usually fitted with shotgun microphones mounted in large windshields. The boom can be tracked over fairly level ground but tracks may be needed in rough terrain.

Camera mounted microphones

In some circumstances it can be a useful ploy to mount a shotgun microphone on a camera so that the two are aimed together without need of a separate sound operator. This does not work when the camera is using a long-focus lens at a distance from the artist, but is an excellent way of picking up effects noises and crowd scenes etc. Some domestic-type cameras are provided with variable-directively microphones coupled so as to 'zoom' with the lens. The effectiveness of this arrangement can be severely limited by the acoustic situation (see page 36).

Radio microphones

Undoubtedly the most useful tool for outside broadcast work is the radio microphone. Within the constraints mentioned on page 72 this gives the artist complete mobility. The microphone/transmitter arrangement can also be useful for static microphones to eliminate a difficult cable run, e.g. for effects microphones at the jumps at racecourses.

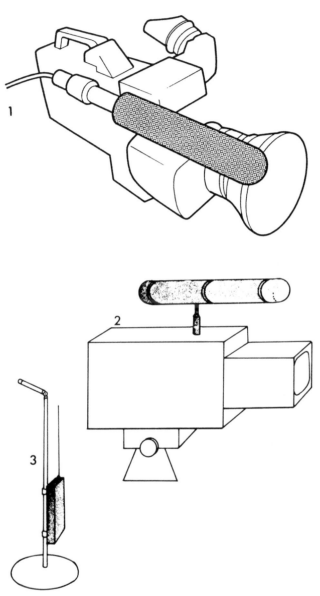

1. A narrow-angle (short-shotgun) microphone with foam windscreen mounted on a lightweight camera. It can be useful for close-working or picking up atmosphere sounds and effects.

2. Shotgun microphone with windshield attached to large professional camera.

3. Small microphone and stand equipped with transmitter for relaying sound from awkward locations.

165

*Not all the action noises made by the actors sound realistic;
some have to be augmented by 'noises off'.*

Sound Effects in the Studio

Most of the noises necessary to give reality to the dramatic scene occur naturally with the movement of the artists, unlike radio drama or most film dubbing theatre practice where the actors stand still and noises such as doors opening, people walking etc. are made by effects operators.

Unreal 'props'

Some of the settings and props are not made out of the material they are supposed to represent and therefore make the wrong noises, e.g. safe doors and 'iron girders' are made out of wood, and rocks out of plastic (possibly polystyrene). When the action would result in such objects making a noise it is necessary to use recordings of the real thing which have to be timed to match the action very accurately.

Offstage sounds

Some effects sounds have to occur out of vision, such as a distant door slam etc. This requirement should be foreseen in the planning stages and a special effects door (or other prop) obtained and placed at the right distance from the dialogue microphone to give the required distant effect. It is worth noting that many of the threatrical conventions for 'noises off' such as shaking metal plates for thunder or dropping piles of scene braces for crashes rely upon their loudness for their effectiveness. When relayed at low volume through a television set they sound just what they are. It is much better to use recordings of real noises.

Bells

Quite often the action calls for the ringing of a bell or buzzer on cue. For the purpose the assistant floor manager can be provided with a bell box — a portable device containing a number of bells and a buzzer with pushbuttons and a battery. The position of the bell box must be set in relation to the dialogue microphone to achieve the right acoustic effect.

Telephones

The best way to make a telephone ring sound natural is to ring the actual phone on the set. This is obviously the right sound and it can be given the right acoustic relationship to the dialogue. The telephone can be rung by means of a special telephone ringer. This gives it the proper mark/space ratio (burr-burr) but arranges that it always starts the first ring instantly on cue. The telephone ringer is also connected to any other telephones used in the action and provides the necessary polarisation so that the actors can hear each other over the telephone when they are working in different sets. The device can also simulate manual or foreign ringing and provides an output of dialling tone.

166

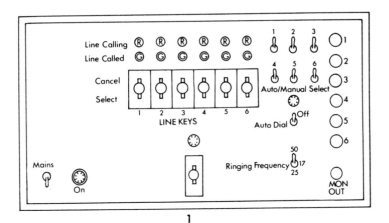

1

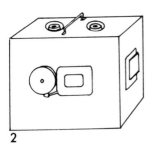

2

1. Unit for interconnecting telephone in the studio. It acts like a six-line manual/automatic telephone exchange (the switches at the top right are depressed for telephones with a dial). Telephones to be rung are selected on the 'line called' key (green light on). If the auto dial key is selected, dial telephones can ring the selected telephone by the action of dialling. Picking up the telephone gives dialling tone, which can be used for the broadcast.

2. Typical bell and buzzer box containing batteries and worked by push buttons.

Discs are used extensively in television for music and sound effects.

Cueing-in Discs

Whenever discs are required for a programme they will have to be cued in with great accuracy.

Analog discs (gramophone records)

In the case of analog discs it is unlikely that an accurate starting point can be located by simply dropping the pickup into the groove, due to the spiral nature of the groove, possible eccentricity of the centre hole and a margin of error of one whole revolution.

To obtain an accurate start it is necessary to play the record ahead of the cue while listening on pre-fade. When the cue is heard the turntable is stopped and rotated slowly backwards by whatever angle is required for it to get up to speed (typically about ¼ turn). Some turntables provide the facility to counter-rotate automatically. The pickup cartridge must be chosen to stand up to this backward rotation without damage.

Quick-start turntables

To achieve accurate cueing by the above method it is necessary to have a turntable that comes up to speed very quickly, otherwise every start will be marred by 'wow', especially in the case of music. There are two basic methods of obtaining a quick start: mechanical and electronic. The mechanical system consists essentially of two turntables arranged one above the other. The lower turntable, which has a heavy flywheel action, rotates continuously and can be raised or lowered by means of a lever or solenoid actuated by a contact on the backstop of a fader. The upper turntable, the platter, holds the disc; it is undriven and at a fixed height. When the cue is required the channel is faded up, the solenoid raises the turntable to contact the platter which takes up the speed very quickly. Alternatively, some quartz-locked servo-operated turntables have high-torque motors and provide very rapid start-up and speed stability (well within ¼ revolution) and are very suitable for cueing purposes.

Slip mats

A cheap method of cueing, provided that a turntable with a reasonably powerful drive is available, is the slip mat. This is a piece of material with a suitable friction coefficient which is placed between the platter and the disc. The turntable is set in motion and the mat held still with the fingers until the cue. When released it will start up almost instantaneously.

Compact discs

These have a built-in cueing mechanism in the form of a time code which describes the position on the track from the start to within 1/25 second. This enables automatic selection of any part of the track. Professional CD machines provide the ability to start precisely on cue (not just the beginning of the track). The cue point can be found by 'rocking' back and forth with a search wheel.

168

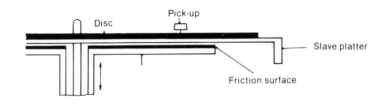

1

Disc
Pick-up
Slave platter
Friction surface

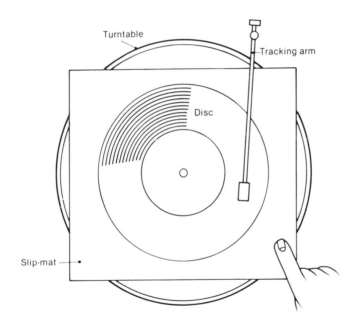

Turntable
Tracking arm
Disc
Slip-mat

2

1. Quick-start professional turntable. The turntable rotates continuously and is raised to friction-drive the record platter. Some quick-start turntables have the facility to rotate backwards to set the cue, with optional muting until fully up to speed.

2. The slip mat fast-start technique.

Microphone positions for balancing a piano depend on the quality of the instrument and the type of music being played.

Televising the Piano

Tone production

The piano is a percussion instrument. When a key is depressed a felt covered hammer hits a string and immediately springs away, allowing the string to vibrate. The string vibrates at its fundamental frequency determined by its weight, length and tension, and also at a number of harmonic frequencies which are multiples of the fundamental note. Some of these are discordant with the fundamental and have to be suppressed. This is achieved by giving the hammers a fairly broad surface and covering them with soft felt and arranging for them to hit the string at a point that damps the most discordant harmonics, notably the seventh and the very high 'clang' tones. If the felt has hardened or the keys are struck very hard so that the felt becomes compressed the upper harmonics are excited and the piano has a harsh tone.

Energy distribution

The treble strings are so thin that they make little sound on their own. They are made to pass over a bridge which transmits the vibrations to a sound board from which most of the tone is radiated.

The sound board is less effective for the very high harmonics which promote brilliance and attack. The nearer the microphone is to the strings the more brilliant is the tone. The bass strings, being thicker, radiate quite strongly at right angles to their length. One way of discriminating against a too heavy bass in a 'distant' balance, is to place the microphone toward the tail end of the piano, looking toward the keyboard.

Choice of microphone position

The microphone position for a piano is determined by:
1. The ratio of attack to warmth of tone required to suit the music and character of the performance.
2. The tone of the instrument and its internal balance.
3. The acoustic conditions and possibly the need for separation from other instruments or noise.

The lid should always be open or off, otherwise the piano will produce a rather 'stifled' sound. In general the best results are obtained by positioning the microphone so that it can 'see' all the strings at a distance determined by the ratio of attack to warmth of tone required. This can usually be achieved along a line running at about 45° in each plane from the strings.

Generally speaking the harsher the tone of the piano and the more percussive the performance the greater should be the distance.

The optimum position can best be found by 'searching' with the microphone on a boom during rehearsal. When this has been found a neat slung microphone can be substituted.

170

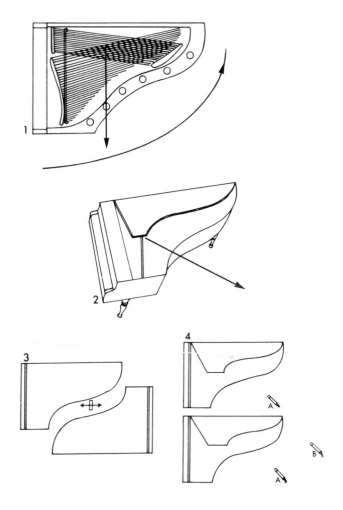

1. A concert grand piano can have considerable bass radiation at right-angles to the strings. A more even balance can often be obtained (with a microphone at about four metres distance) by moving it toward the tail.

2. Best results are usually obtained with a microphone on a line diagonal to its three major dimensions so that it can 'see' all the strings.

3. Two pianos 'wedded' for a duet. The lids have to be removed for the arrangement.

4. Two pianos in line formation with balance for two microphones AA, or one B.

171

The rhythm piano requires a very close microphone at the treble end of the instrument.

Televising the Rhythm Piano

The technique for broadcasting the piano in light music and dance bands is quite different from that used for classical music.

The piano in a rhythm band

Traditionally the piano was part of the rhythm section of the dance band, its function being mainly to provide rhythm chords, filling-in phrases (mostly in the treble) and occasionally the melody. Modern orchestration tends to be rather more complex but the requirement remains for a very crisp brittle tone (particularly at the top end) and good separation from the rest of the band so that, even when played quietly, it can be heard in competition with the much louder brass and saxophone sections.

Microphone techniques

The necessary tone is best achieved by using a unidirectional (cardioid) microphone pointing either almost vertically toward the top strings at a distance of about 20 cm (8 in) or pointing straight down into one of the holes in the iron frame (normally the second from the treble end) and only slightly above the frame so that it looks straight at the soundboard. This latter position does not produce quite such a crisp tone as the microphone pointing at the treble strings, but it can produce a remarkably well balanced result. The reason for this is probably due to the fact that, although the sound board is designed to radiate all frequencies in phase, there is inevitably some antiphase cancellation at the upper end of the scale where the wavelengths are small. Concentrating on a small area tends to eliminate this effect.

A very neat arrangement can be made for either application by using an electrostatic microphone with short extension tube as a table stand resting on a foam rubber pad inside the piano.

The best mic position for most upright pianos is with the top lid or front cover off pointing at the top strings. If this is objectionable in shot the microphone can be placed behind the piano, pointing towards the top end of the soundboard at about 20 cm (8 in) distance.

Multitone and electronic pianos

Some upright pianos have a third pedal (in addition to the sustaining and soft pedals) which causes a set of little cloth strips fitted with metal ends to interpose between the hammers and the strings. This greatly increases the harmonic content of the sound and produces a metallic tone, slightly like a harpsichord. Multitone pianos have been superseded by electronic keyboards with which a wide variety of tone colours can be obtained. These are usually picked up by a microphone placed in front of their cabinet (loudspeaker), or by direct injection into the sound mixer from a DI box.

172

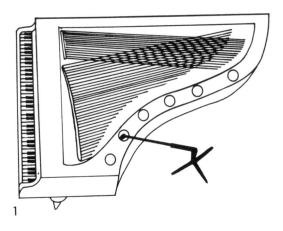

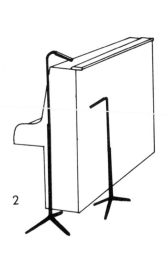

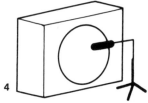

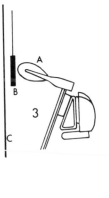

1. The microphone head would be just above the lip of the hole in the frame. A neater arrangement can be made by resting a short stand on the piano frame or clipping it inside the case.

2. The best arrangement is with the microphone over the open lid at the treble end. Alternatively, it can be placed behind the sound board.

3. *Arrangement for multitone piano.* A, felted hammer. B, cloth tab with metal end. C, piano strings.

4. With an electronic keyboard it is merely necessary to position a microphone (usually cardioid) in front of the loudspeaker, or take a direct output if available.

Various Keyboard Instruments

Apart from the piano there are several keyboard instruments likely to be encountered in television productions.

Harpsichord

The harpsichord is rather like the piano except that the strings are plucked with quills instead of struck with hammers. This produces a small tone that is very rich in harmonics. These harmonics are largely centred on the frequencies where the ear is particularly sensitive, so the harpsichord sounds much louder than a programme meter measuring its output would lead one to believe.

A particular feature of the harpsichord is its loud action noise, which stems directly from the plucking action.

The best method of broadcasting the harpsichord is usually to employ two microphones, one underneath pointing towards the soundboard, but angled to discriminate against any noise from the pedals, and another microphone above in a similar but rather nearer position than that normally used for a classical grand piano balance.

Clavichord

The clavichord is a smaller, simpler precursor of the piano. Normally in the shape of a rectangular box, it sits on a table. The strings are struck with a brass tangent which remains in contact with them while the key is depressed so that vibrato (small rhythmic changes of pitch) can be applied by varying the pressure on the keys with the fingers. The tone produced is very small and delicate and calls for a very close balance unless the instrument is completely solo; this is best achieved with a microphone overhead and to the right.

Celeste

The celeste is a keyboard instrument in which the tone is produced by bells. Most of the sound comes out of the back, which has an open fret covered with thin fabric.

Celestes vary considerably in purity of tone and power. They usually sound best with the microphone about two metres (six feet) away from and slightly above the back, but the need for separation from other instruments usually dictates a much closer balance, often about 20 cm (8 in) from the back.

Organ

Most of the organs encountered in television productions are electric. Their sound comes out of loudspeakers either contained within the console or in separate units. It is possible in many cases to take a direct feed from the organ to the sound mixing desk but, unless the organ is properly adapted for this purpose, there may be matching problems and it is simpler to place a microphone in front of the loudspeaker.

174

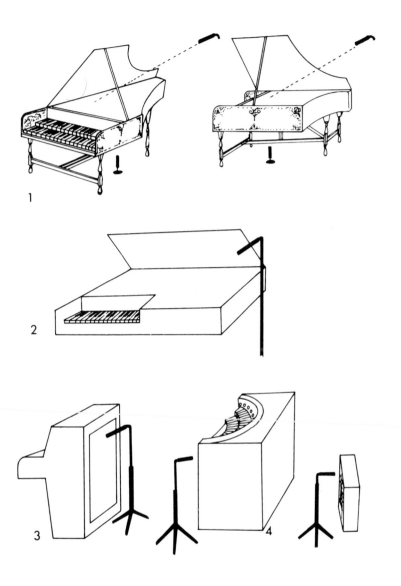

1. Two views of a harpsichord showing microphone positions under and over the instrument.

2. Clavicord, showing suitable microphone position.

3. Celeste, with microphone at the back.

4. Electronic organ with alternative position for microphone for internal loudspeaker and external loudspeaker.

Stringed instruments sound best with the microphone roughly perpendicular to their front faces.

Strings

The 'family' of strings used in most orchestras consists of violins, violas, cellos and double basses.

Violin

Most of the tone of the violin is radiated from the front face and from the 'f' holes. The back, which is connected to the front by a peg called the sound post, also makes a powerful contribution to the tone.

The high harmonics which give brilliance to the tone radiate in a narrow lobe at right angles to the face of the instrument.

If a fairly distant microphone technique is used, say two metres (six feet) or over, the best sound is produced along this line. If, however, a very close technique is necessary to give separation from other instruments it may be better to work somewhat off axis, otherwise the tone can be rather 'whiskery' and variable if the violinist sways about.

The violins in an orchestra sit in pairs (called desks) facing the conductor, sharing a music stand. They are usually divided into two sections of 'firsts' and 'seconds' or, in some types of light music, three groups, A, B and C.

Viola

The viola is a larger version of the violin. It is tuned a musical fifth lower. It has thicker strings and produces a less brilliant tone which is not quite as directional as the violin.

Cello

The cello is tuned an octave lower than the violin. The strings are not unduly thick in relation to its size and it is fairly directional in its upper register. Cellos, or more properly, in the plural, celli, are also arranged in pairs in the orchestra, sharing a music stand between two.

Double bass

As played in the 'straight' orchestra the double bass does not constitute much of a problem as it is not very directional, but it is sometimes necessary to use a fairly close microphone technique to obtain clarity.

Rostrum resonance

Both the cello and the double bass are connected to the floor by the spike on which they rest. This means that the floor acts as an extension of the instrument, and if it is a rostrum great care is necessary to ensure that it does not resonate. This is achieved by making a false top with hardboard laid on felt and providing stiff bracing inside. In general the basses sound better on the studio floor.

1. The violin radiates its high tones and harmonics in a fairly narrow angle slightly forward of normal to the front face. The low frequencies occupy a much broader field.

2. General appearance of violin showing position of sound post and 'f' holes. The sweetest tone (from within the resonating chamber) comes out of these holes.

3. One 'desk' of violins or violas occupies a rostrum space of approximately 1.8 × 1.3 metres (6 × 4 ft).

4. Desk of celli needs at least 2 × 1.8 metres (6.5 × 6 ft) unless the musicians sit diagonally. If possible 'cello boards', A, should be used.

5. When basses are used on rostra care must be taken to avoid resonance. The rostrum must be well braced and, if possible, a 'sandwich' of felt and extra top section should be used. A pair of double basses occupies a space of about 2 × 1.5 metres (6.5 × 5 ft).

177

Woodwind Instruments

In deciding where to place microphones for the woodwind there are two different approaches according to whether the music is 'straight', i.e. orchestrated for a natural balance in the concert hall, or designed for microphone assistance with little if any attempt at internal balance. Most modern and all 'pop' music comes into the latter category.

Flute and piccolo

The flute and piccolo produce a very pure tone when heard from a distance, i.e. the microphone at least two metres (six feet) away. At close range they have a 'breathy' quality which is often exploited in popular music to produce a crisp effect. In the ultimate the microphone is placed a few centimetres from the mouthpiece so that the breath puffs give each note a sharp leading edge. A special microphone is available which fits into the end of the flute for extra separation.

Oboe

The oboe has a sweet but piercing tone due to the fact that it is rich in harmonics, most of which are centred on frequencies for which the ear is particularly sensitive. The sound that emerges in line with the bell is particularly strident but this is directed toward the floor. Microphone distances of less than one metre (three feet) are seldom used. The instrument sounds best from a considerable distance. The same is true of the cor anglais (or English horn). This is a larger instrument with a natural key of F, a fourth lower than the oboe. It has a warm complex tone often used to suggest a 'pastoral' atmosphere in programme music. This is created by the unusual shape of the bell, which is slightly spherical and closed in somewhat at the end.

Clarinet

The clarinet has a variety of different qualities according to the register in which it is played and the way it is blown. In the high register it can be very piercing. It can also be played 'sub tone' (below its fundamental register), when it sounds very soft and requires a very close microphone technique.

Because of its many registers (usually divided into 'medium', 'chalumeau', 'acute' and 'high') it is difficult to play across the breaks between. For this reason clarinets are made in several pitches, the ones in most common use being B flat and A. Military bands often contain an E flat clarinet which has such a piercing tone it is capable of cutting through the rest of the band.

Bassoon

The bassoons are normally only used in straight orchestras. Their sound is not very directional so they pose few balance problems.

178

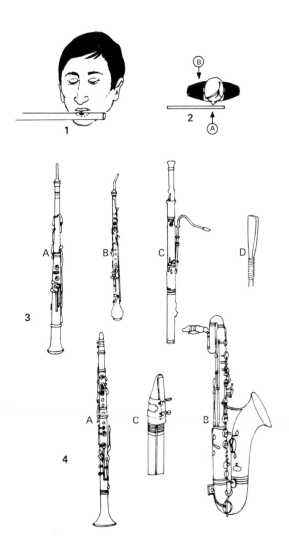

1. The tone of the flute is produced by blowing over a small aperture. (It is said that 'you don't play the flute, you stand beside it and pop the tune into it'.)

2. A crisp tone is produced with a microphone in front of the mouthpiece (position A; a purer tone from position B).

3. A, oboe, B, cor anglais and C, bassoon use double reeds. D, double reeds are two reeds fixed parallel on either side of the mouth tube, leaving a narrow gap between.

4. A, clarinets and B, saxophones use single reeds. C, single reeds form a narrow gap with the mouthpiece.

Brass Instruments

The brass instruments require a different approach from the woodwind by virtue of their greater power and directional properties.

Trumpets and trombones

Trumpets and trombones project their sound most powerfully in line with their bells. The high harmonic overtones that give brilliance are confined to quite a narrow angle so that if the microphone is right on axis the sound quality varies considerably if the player changes direction. It is, therefore, invariably better to work slightly over the direct line.

When planning the layout for trombones it is necessary to allow space for the extension of the slides. It is also worth considering that the trombones make a spectacular effect when viewed from the side with their slides moving in unison, so they should be arrnaged in a straight line that is accessible to the cameras.

Mutes

Various mutes are available for the brass. These alter the tone and reduce the volume in various degrees. In general the straight conical mutes produce a hard tone that is not very much less in volume than the open bell (if a microphone is not in direct line). The cup mutes that cover the bell reduce the volume very considerably and produce a softer tone. All mutes, especially the cups, tend to make the instrument much less directional so that it is satisfactory to work across the microphone.

Horns

The horns require special care in positioning because they point backward almost directly behind the performer so that the surface immediately behind plays an important part in the production of tone. It should provide reflection, but no resonance, and must not be so close that the performer is inhibited from playing at the proper volume.

Euphoniums and tubas

Euphoniums and tubas are mainly used in brass bands and are not very directional; they tend to sound best from an overhead position.

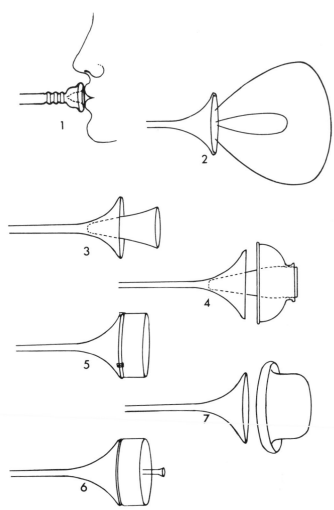

1. In the brass instruments the tone is produced by pressing the lips against a cup-shaped mouthpiece so that they form a pair of reeds, the air flow being controlled by the tongue.

2. Because of the shape, most brass instruments, particularly the trumpet, produce their most brilliant tone in direct line with their bells (the narrow lobe represents the HF response). Various mutes are used with trumpets and trombines (and to a lesser extent horns).

3. The straight mute which produces a hard tone suggesting distance.

4. The cup mute — a soft smooth tone.

5. The hush mute — a can with absorbent material inside clips over the bell to produce a very soft tone.

6. The Harman or 'wow wow' mute for shrill 'talking' effect (made by covering the end with the hand).

7. Bowler hat for swell effect.

Small classical orchestras can often be balanced with one microphone.

Chamber Orchestra

The term chamber orchestra suggests a small combination of classical musicians playing music that was written to be played in fairly intimate surroundings. It may be a quartet or a small orchestra of the type for which much of Mozart's work was written. In any case the chances are that there will be a conventional layout which gives good internal balance.

Pictorial presentation

In keeping with the type of music, chamber groups and small classical ensembles are often depicted on television in realistic period settings so that it would be incongruous to see microphones in shot. Luckily the internal balance is such that in many cases only one microphone (or a pair if the sound is in stereo) is required. It can be at quite a distance (say about three metres or 10 feet), where it could easily be hidden behind a chandelier in a long shot.

String quartet

The string quartet has a conventional layout designed to help the musicians to play in concert and balance with each other.

In this type of music the parts are woven together in such a way that sometimes all make an equal contribution to the texture and sometimes each comes into prominence in turn. It should be possible to find a position for the microphone, usually about two to three metres (six to ten feet) in front and above, where these conditions are met. If a piano or harpsichord is added another mic will be required. If the sound is in stereo a stereo pair of microphones is required in the main position.

Classical orchestra

The type of orchestra used by most of the great composers up to about 1800 consisted of about 22 strings in the form 8-6-4-2-2 (1st violins, 2nd violins, violas, celli and basses), double woodwind, i.e. two of each (flutes, oboes, clarinets and bassoons), four or five brass (two trumpets and three trombones), two or four horns, and timpani.

The standard layout, comprising a semicircle, with the strings in front, woodwind and finally brass and percussion behind, makes a logical combination that should balance on one microphone (two if stereo) somewhere above and behind the conductor's head. However, it is usually necessary to add at least one other microphone to accentuate the woodwind and give extra clarity and better perspective.

The procedure is to experiment to find the best position for balance and at the same time the best compromise between clarity (nearness) and warmth of tone (reverberation), then to listen critically to the rehearsal and tell the conductor of any points where the balance is not satisfactory. He will expect this sort of guidance and can usually make the necessary correction.

182

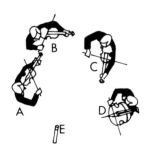

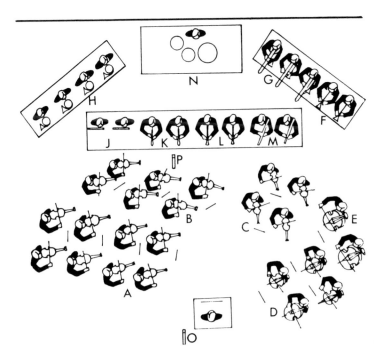

1. The string quartet. A, first violin. B, second violin. C, viola. D, cello. E, typical position for microphone at about 2–3 metres (6–10 ft) high.

2. The classical orchestra. A, first violins. B, second violins. C, violas. D, celli. E, double basses. F, trumpets. G, trombones. H, horns. J, flutes. K, oboes. L, clarinets. M, bassoons. N, tympani and percussion. O, possible position for a single microphone balance. P, position for woodwind accentuation microphone about 2–3 metres (6–10 ft) high. For stereo the main microphone would be a stereo pair augmented by at least two others 'pan-potted' into position.

The symphony orchestra generally needs several microphones to provide the clarity and attack needed for television.

Symphony Orchestra

The modern symphony orchestra can employ up to about 100 musicians, according to the music to be played. A typical combination would consist of about 66 strings in the form of 20-16-12-10-8 (using the notation explained on page 182), triple woodwind, seven brass, including bass trombone and tuba, four horns and at least two percussion players, two harps and possibly a celeste.

The factor of size
One of the main problems in televising a symphony orchestra is its sheer size. Most symphonic music, with the exception of some modern pieces, is written to produce an internal balance within the orchestra but this only really works in terms of balance, perspective and acoustic effect from a position well back in the hall.

Theoretically, given ideal acoustic conditions and layout, it should be possible to balance the whole orchestra on one microphone (or stereo pair) with the conductor correcting points of balance with the guidance of the sound supervisor, and some excellent recordings have been made in this way. In practice though, it is usually necessary to add accentuation microphones for the various sections (particularly the woodwind) which, in the case of stereo, would be 'panned' into the proper geographical relationship. The general approach is to create the aural equivalent of a wide-angle long shot.

Television perspective
Unfortunately when televising orchestras, directors do not stick to overall long shots because the figures of the musicians would be too small for the viewer to direct his gaze around the orchestra as one does in the concert hall. Instead they track or zoom the cameras in to follow the predominant musical line in close-up. This faces the sound supervisor with a dilemma. If he maintains a distant flat perspective the sound is incongruous with the close-up pictures and produces a frustrating effect, but to follow the visual presentation with matching close up sound would be musically disastrous.

Orchestral balance for television
The solution is to meet the situation half way: to balance the orchestra on a number of microphones, each rather closer than would be required for a sound-only recording. To these is added an overall stereo pair and, given good acoustic conditions, a further stereo pair well back in the hall. If necessary artificial reverberation (suitably delayed) is added, especially to the 'spot' microphones. This gives the overall sound a clear attack which matches well with close-up pictures without the necessity to adjust balance for pictorial reasons.

184

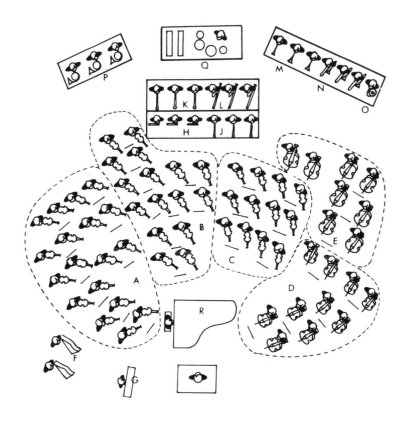

Typical 'straight' layout for full symphony orchestra. A, first violins. B, second violins. C, violas. D, celli. E, double basses. F, harp. G, celeste. H, flutes. J, oboes. K, clarinets. L, bassoons. M, trumpets. N, trombones. O, tuba. P, horn. Q, tympani and percussion. R, piano if required.

Some conductors favour variations of this arrangement, notably to exchange the second violins' position for the celli. For television, spaced layouts are often employed to give a better opportunity to take pictures of the soloists in the orchestra (especially the woodwind) without giving too much emphasis to the legs of the person sitting behind.

Featured showbands tend to play stylised orchestrations requiring specialised multi-microphone technique.

Concert Orchestra

Some television productions employ very large orchestras of almost symphonic proportions playing stylised orchestrations that require multi-microphone technique to achieve balance.

These may be used as accompanying orchestras for prestige light entertainment productions involving star vocalists or instrumentalists or as concert performers in their own right. In either case the balance technique is dictated not so much by the lack of internal balance in instrumentation (in fact some featured orchestras employ as many string players as a symphony orchestra) but by the type of orchestrations used, the characteristic quality of the required sound and possibly the versatility to cope with different types of accompaniment. In addition to a large string section the orchestra will probably have a complete dance band section with five saxophones and eight brass, horn, double woodwind, piano, harp, keyboards with synthesisers, a large rhythm and percussion section including electric bass and guitars and possibly drum synthesisers (electronic drums). Each of these will require a microphone in front of its respective loudspeaker and may also provide an individual or mixed output for direct injection into the sound mixer.

String treatment
Despite the volume of strong tone available, many microphones are used, one for about every four desks of instruments (more if the layout is widely spread). This is to ensure a really solid string tone with good clarity and separation from the other sections so that the desired amount of artificial reverberation and possibly frequency response shaping can be applied individually. All the other instruments have practically one microphone for each.

Brass, saxes and rhythm section
The treatment for these is as described on page 192. It is important that the rhythm drums occupy a central location close to the brass.

Visual considerations
If the band is featured in vision, it is necessary to provide good access for camera tracks around the sections including side shots along the rows. All solo instruments such as woodwind, accordion and piano must have good visual as well as aural separation to enable clean close-ups to be taken.

Foldback
If the soloist is separated from the orchestra by more than a few metres it will be necessary to provide foldback of the rhythm section as well as the soloist's own sound on a stage monitor speaker.

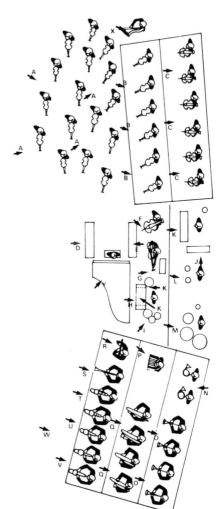

The microphones (arrows) are the minimum likely to be required. Each of the brass players may require his own microphone and the drum kit could require several more, including two for the snare drum.

The microphones cover the following sections: A, violins. B, violas. C, cellos. D, E, keyboards and synthesisers. F, bass. G, guitar. H, bass drum. I, hi hat. J, vibraphone. K, percussion. L, bongoes. M, tymps. N, horns. O, trumpets. P, accordion. Q, trombones. R, flute. S, oboe. T, baritone sax. U, alto saxes. V, tenor sax. W, overall microphone for band section. X, harp. Y, piano.

The Light Orchestra

Light orchestras come in all sizes and combinations and play all types of music from light classical to 'pop'. It is, therefore, not possible to generalise as regards their balance technique except in one respect: they are all likely to require a multi-microphone balance, particularly on television.

Light music balance

A typical combination to play straightforward light music would consist of about 22 strings (typically 8 first violins, 6 second violins, 4 violas, 2 celli and 1 or 2 basses), 3 trumpets, 2 or 3 trombones, 2 horns, double woodwind (i.e. 2 of each: flute, oboe, clarinet and bassoon), some of whom may also double on saxophone, piano, possibly harp, tympani and percussion.

Even if the orchestrations are written for internal balance this is unlikely to be achievable on one microphone. It will be seen that the power of the wind, brass and percussion is very similar to the symphony orchestra but the string section is only half its size. It therefore becomes necessary for the microphone to close in on the strings to achieve balance. As a result the strings will sound much nearer than the rest of the orchestra so we have to add equally close microphones to the other sections, fading up just enough of their output to restore perspective while maintaining balance.

Once it becomes necessary to use a close microphone no section must be left without one or it will be missing, or sound very distant.

Multi-microphone techniques

The use of separate microphones for each section provides several advantages:

1. Time is saved in finding the optimum position for a single microphone. This can be a premium in the television rehearsal situation.
2. The music arranger is freed from the requirement to consider internal balance and can write for all sorts of novel effects that exploit microphone techniques.
3. Greater freedom of layout. The orchestra can be arranged in a spread formation to help separation or for the sake of appearances.
4. The use of close microphones with added artificial reverberation produces a 'crisp', brilliant tone which suits modern orchestration and is congruous with close-up shots of the musicians.
5. Various forms of enhancement, e.g. tonal modification, can be applied individually.

Spread-out split layout for visual effect. Because of the unnatural spread of the instruments, and possibly also to balance for modern orchestrations, a large number of microphones, each at fairly close range, is required.

The microphones (pictured here as arrows) accommodate the following sections: Microphones A, B, C, D, violins. E, violas, F, celli. G, celeste. H, piano. J, acoustic guitar. K, double bass. L, electric guitar amplifier. M, tympani. N, bass drum. O, side percussion. P, vibraphone. Q, flute. R, oboes. S, clarinets. T, bassoons. U, trombones. V, trumpets. W, horns (they project their sound backward). X, overall string microphone. Y, overall woodwind and brass microphone. Z, bongoes.

X and Y can be very useful to restore cohesion which may be lost with very close microphone technique. They also tend to bridge the gap between the direct sound and the added reverberation when an acoustically dead television studio is used.

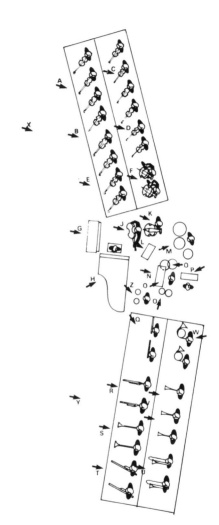

Opera on television can be shot as a theatre relay or, realistically, as a drama with music.

Live Opera and Large-Scale Musicals

Televising full scale operas, or even excerpts from operas, can present interesting problems from the sound point of view.

Opera on television is shot largely in close-up with realistic settings. It can play to an audience of many millions most of whom will expect to hear the words. The problem is to obtain clear diction from mobile singers without so subduing the accompaniment that the musical effect is lost. Stereo sound can help, by placing the soloists centre-stage with the accompaniment to the sides.

Studio opera

The larger operas require an orchestra of almost symphonic size and large sets to include many long shots, so the major problems are space separation and time-lag.

The space required to accommodate a large orchestra and even one spectacular set would be enormous. Even if such space were available the power of the orchestra would place restriction on the working distance of the booms and restrict the shots. The orchestral sound reaching the booms, and the singers, would be unbalanced, variable, and come in late, and the singers and orchestra conductor would be unlikely to be able to see each other.

Separate orchestral studio

The best arrangement is to put the orchestra in a separate studio from the action and feed each to the other via loudspeakers. The singers hear the orchestra via column loudspeakers mounted on and tracked with the booms. Then the orchestral sound is always effectively close to them, localised to the back of their microphone and balanced so that orchestral pick-up on the vocal microphone is not so serious. Loudspeakers are also placed strategically in the orchestral studio to relay the singers to the orchestra so that aurally they are close to each other and time lag is eliminated. The singers see the conductor via a camera and monitors scattered about the sets. A répétiteur (deputy conductor) is employed to watch the monitor and relay the beat. With this arrangement the booms can work at a distance of about two metres (six feet) and a smooth effect is obtained.

Theatre relays of opera

When opera is relayed from the opera house there is less control of the situation. The vocal pick-up is by microphones along the footlights and slung behind the proscenium arch. The rather distant sound is mitigated by the more theatrical presentation and the fact that the action tends to work towards the audience.

A major problem is obtaining a good balance from the orchestra in the pit. This is achieved by placing microphones within the orchestra and adding their output to a microphone slung high over the top.

190

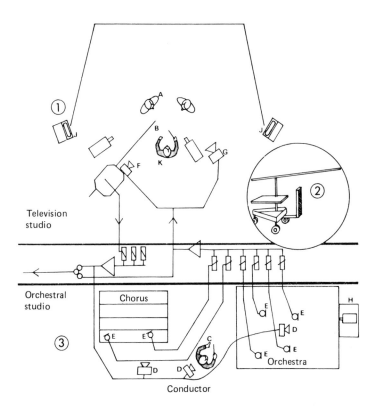

Set-up for remote orchestra production of opera. Singers (A) in television studio (1) are picked up on boom (B) for broadcast and also relayed to orchestral studio (3) where they are heard by the conductor (C), orchestra and chorus on loudspeakers (D). The chorus and orchestral sound is picked up on microphones (E) for broadcast and also fed to artists on special boom-mounted loudspeaker (F) (see inset 2) and mobile loudspeaker (G). The camera (H) gives a picture of the conductor for slung monitors (J). These are arranged to be on the artists' eye line or the beat is relayed by the répétiteur (K).

Big (Dance) Band

This type of band consists of a brass section (the big bands have 4 trumpets and 4 trombones), a saxophone section (normally 2 altos, 2 tenors and a baritone) and a rhythm section consisting of piano, guitar, bass (string bass or electric bass guitar) and drums.

Rhythm section
The rhythm section is a very important feature of the dance band. In addition to supplying an important element of the sound, which should be reproduced with almost equal volume to the tune, it drives the rest of the band along and gives 'lift' to their performance.

Brass and saxes
The saxes require a microphone each, as do the brass, although it may be possible to manage with one between two and one overall for the brass section.

Rhythm piano
The piano requires very close miking (see page 172) to achieve separation from the band. The pianist could also have several electronic keyboard instruments which play through loudspeakers (cabinets) for which individual mics will be required. Alternatively he may do his own mix with a sub-mixer so that only one cabinet will be used.

Double bass
The bass, whether it be a string bass or electric guitar, should sound almost as loud as the rest of the band. This is simple to achieve with an amplified guitar but much more difficult with a double bass which, when played pizzicato, i.e. plucked, produces a very small sound. It requires a unidirectional microphone within a few centimetres of the strings close to the bridge. Alternatively an omnidirectional microphone can be used softly suspended within the bridge, unless the bass has a transducer pickup, in which case the cabinet will need a mic.

Drums
The drums of a dance band have an enormous volume range. Nevertheless, to be reproduced with proper clarity they require a close microphone technique. See Rock Groups (page 194).

If the drums are on a rostrum it needs to be very firm and acoustically damped to prevent spurious resonances travelling up the microphone stands and blurring the sound.

Guitar
The guitar is almost invariably electrically amplified and a microphone can be placed in front of the loudspeaker cabinet. See also Rock Groups (page 194).

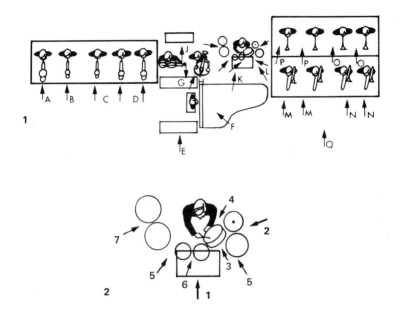

1. *Typical television layout.* Note the rhythm section in the middle to keep the musicians (who are really too separated for ensemble) in time. The microphones (arrowed) cover A, baritone saxophone (sometimes doubling flute). B & C, alto saxes. D, tenor saxes. E, synthesiser and cabinet. F, piano. G, keyboard synthesiser. H, double bass. J, guitar amplifier. K, bass drum. L, side drums and hi hat. M & N, trombones. O & P, trumpets. Q, overall brass microphone (to give cohesion and a sense of power to the brass sound).

2. *Overhead view of drum kit.* A close microphone technique is required to produce a crisp attack. Short-decay reverb can be added to the kick (bass drum) to give a more powerful effect. The mics will include the following. 1, bass drum. 2, hi hat (pedal cymbal). 3 & 4, top and bottom snare (side drum). 5, overheads (for cymbals). 6, tom-toms. 7, bongoes. See also Rock Groups (page 194).

Vocal

The vocalist's mic should be 'folded back' to a monitor loudspeaker on stage in front of the vocal mic and to the rhythm section, who may also require a feed of the vocal on headphones.

Rock Groups

All the instruments in rock groups, with the possible exception of acoustic drums, are amplified through powerful loudspeakers.

Drums

These form the centrepiece of rock music, with a powerful beat that should be prominent in the balance. They can be played acoustically or synthesised or a mixture of both. For good internal separation, the key will typically be covered with two overheads, 5 or more tom-toms, a top and bottom snare, hi-hat and kick (bass drum). The *snare drum* has a set of springs which can be brought into contact with the lower head, prolonging the note and giving it a 'whiskery' quality which can be 'fattened' by adding reverb with a small room acoustic.

Electronic drums can be used to increase the scope of the acoustic kit. Instead of drums they use pressure pads which trigger synthesisers to produce a variety of sounds such as synthesised drum sounds or musical pitches. The pads are velocity sensitive and can trigger different sounds with different levels of impact. The synth outputs can be mixed on stage and may be fed direct to the recording via a DI box.

Guitars

It is usually best to mic the guitar cabinet (loudspeaker) as this output will include all the player's effects pedals and any (intentional) distortion produced by his amplifier and LS. When guitars are used with cascaded pedal effects, or as triggers for synthesisers, where there is a risk of fingering noises becoming intrusive, it may be necessary to use a noise gate.

Bass guitars may be picked up straight from the instrument with a DI box, but will probably need a lot of bass lift in the EQ, to which some of the cabinet sound can be added.

Keyboards

These are used with synthesisers. Often several keyboards with different synths are used so it is best to rely on the musician's sub mix by micing his cabinet. Alternatively, if multitrack recording is possible, separate tracks can be made and mixed down later.

Vocals and acoustic instruments

The sound must be fed back to the musicians on stage on monitor loudspeakers. Microphones with clean off-axis cutoff are required to minimise the chance of howlround. If the vocalist is mobile the microphone/monitor relationship must be checked for all positions.

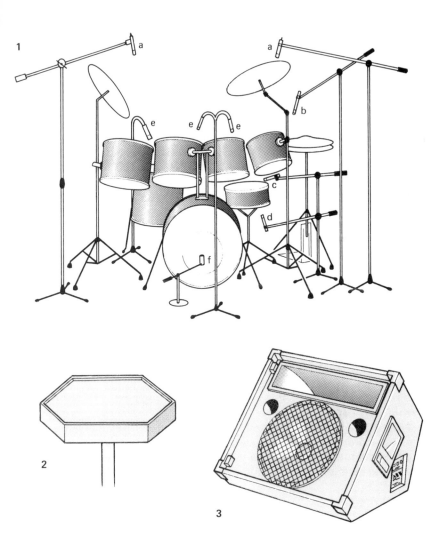

1. Typical drum setup. a, overhead mics. b, hi-hat. c, top snare. d, bottom snare. e, tom-toms. f, kick.

2. Pressure pad to trigger synthesisers.

3. Stage monitor loudspeaker to feed back vocals and acoustic instruments to the musicians.

Post-Production, Sound Dubbing and 'Sweetening'

No matter how carefully a sequence is shot it is most unlikely that the result will be satisfactory or complete in vision or sound without editing and post-production work. There may be occasions, under controlled conditions, when sound and pictures can be edited simultaneously, but this requires carefully matched levels and background sounds for each take. It is surprising how backgrounds can change without its being noticed between discontinuous takes of the same scene, and it is always wise to record some extra atmosphere (wild track) to cover edits.

In many cases it is advisable to deal with the vision and sound separately. As to which should have priority in the editing, it depends upon the programme material. Obviously in the case of live music or speech the sound must be complete and continuous and the vision cut to match.

Simple sound dub

The simplest way to create a sound dub is to copy from one VTR to another, intercepting the sound via an audio mixer between the two. The audio mixer can have other sources available to it, such as music and effects and the ability to shape the frequency response and add reverb etc. There are two basic disadvantages in this arrangement:
1. The process results in a second-generation copy of the video, which can entail loss of quality.
2. If a lot of manipulation is required for the sound this will entail tying up two VTRs through all the attempts to perfect the audio recording, with consequent wear on the VTRs.

Audio lift-off

A better alternative is to lift off the sound from the edited video onto an audio recorder (preferably multi-track), make all the necessary alterations to the audio recording, using a VCR copy if picture matching is important, and then lay the completed audio back onto the original video recording.

Synchronising

The problem with the above arrangement is maintaining synchronism between vision and sound, because even if the video and audio machines are started simultaneously it is unlikely that they will remain in sync for long. So it is necessary to use a synchroniser in conjunction with a time code signal recorded on each machine.

196

Time code

A suitable time code for post-production synchronising is the SMPTE/EBU. This is a ten-digit number code that is recorded along the length of an audio channel or on the cue track of videotape or film. It provides an accurate reference, which acts like sprocket holes, to identify and locate a position on the recording in terms of hours, minutes, seconds, frames and bits of frames. It can relate either to the time of day or to the start of the activity.

The video and time code can be dubbed from the VTR on to a VCR, which can then become the *master* controlling all the other machines (*slaves*) and keeping them in sync. This is achieved by the *synchroniser*, which compares the time code signal from the different machines and on detecting a difference provides a correction signal to keep the machine in perfect synchronism. The time code can be 'burnt in' to the video, i.e. superimposed on the picture so that the exact location of each item can be logged.

Once all the machines are locked together they can be 'rock and rolled', i.e. the tracks can be moved back and forth to add or replace bits of dialogue or effects etc.

Multi-track

The use of multi-track recording is invaluable on post-production work as it makes it possible to lay down a number of different sounds, such as effects and music, in sync with the picture, manipulate them individually and then mix them back onto the original sound track of the VTR.

Digital post-production

Digital memory recorders are available, offering much the same facility as multi-track machines but recording digitally on to Winchester discs. This means that they have instant access to a large number of tracks, some of which can be prerecorded to form a library with consequent saving in assembly time. The output can be triggered manually or by keying from a time code reference number.

Digital recording offers perfect quality through any number of dubs and many sophisticated forms of sweetening, including all types of reverb, and the ability to change the time or pitch of the recorded material (by changing the clocking rate between input and output). This can be a very useful feature for foreign language dubbing to match lip movements with sound.

Tape recorders are used to play in specially-recorded music and effects and for supplying the sound for mime.

Tape Recording

Quarter-inch tape recorders are used extensively in the production of television sound. The most useful type, suitable for most purposes, is the twin track 7½/15 in/s (19/38 cm/s) type. Twin tracks are useful for a number of television operations, such as mime. When artists mime or dance to pre-recorded music it can be helpful to provide two different balances, one for broadcast and the other with a particular emphasis on rhythm to help the artist to synchronise.

Sometimes vocalists are pre-recorded (for practical reasons) and the orchestral accompaniment has to be added live. In this case the second track can contain a rhythm beat and bar count fed to the conductor to ensure that the orchestra remains in synchronism through long breaks in the vocal line.

Another use for twin tracks is when effects recordings have to have different characteristics while maintaining the rhythm. An example is shots of a vehicle seen alternatively from inside and outside. The two different-sounding, but related, effects are recorded one on each track, and the sound is cut between them in synchronism with the shots.

Incidental music recording

A good deal of incidental music and music for mime is recorded initially on multi-track recorders. It is possible to synchronise a number of tape recorders with each other and with a VTR by means of a digital time code, using computer technologies. SMPTE/EBU time code can be used, or a system called 'Maglink' which employs an analog-derived frequency-shift code which does not create cross-talk problems.

Multi-track recorders are made with 8, 16, 24 or 32 tracks. This enables each section of the orchestra and the vocal to be recorded simultaneously but separately so that different treatments can be applied to each, and the mixing to what is called the 'reduction copy' can take place without having to extend an expensive orchestral session. The reduction copy can be a single track recorded across both tracks or have vocal on one track and accompaniment on the other. This enables small adjustments to be made to the vocal sound to match the picture.

When music or effects have to be fitted to visual action it is not always possible to judge the length of material required. It is a good scheme to use two tape recorders for reproduction, arranging that the sound sequences are on the long side and spliced on to alternate reels so that the next cue can be set up while the other machine is playing. Often long continuous backgrounds are called for in addition, in which case there will be a need for a third machine. Continuous-loop cassette machines can be useful for this purpose, but the loop must be sufficiently long to prevent the repetition becoming obvious.

1

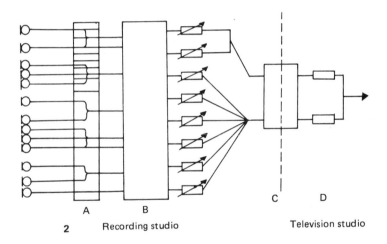

2 Recording studio   Television studio

1. *Full track and multi-track.* The full-track machine uses the whole width of the tape for one track. Although this gives the best signal-to-noise ratio, few recordings are now so made in television. Multi-track machines use two or more tracks separated by 'guard bands' i.e. small spaces with no modulation. In stereo working it is normal to allow a slightly wider guard band on repro than on record to reduce crosstalk and improve the signal/noise ratio when the recording is played 'mono'.

2. *Transmitting multi-track recording.* A multi-microphone balance can be mixed (A) to record on 8 (or 16) channels at the recording studio (B), re-mixed (reduced) to 2 channels (C) and finally in the television studio (D) to one channel for transmission.

Quarter-inch sound tape can be edited with great accuracy by cutting and splicing.

Tape Editing

Most professional use of tape machines involves editing the tape. This may be to assemble the material into order, adjust its length, or merely to add 'leaders' (blank tape of different colours) to identify start and stop cues.

Cutting the tape

Editing ¼ inch tape is achieved by physically cutting the tape. The procedure is to play back the tape until the cue is heard. The machine is then stopped and the two spools rotated slowly backwards and forwards by hand until the exact point of the cue is established. A mark is then made on the tape to coincide with the reproducing head (normally the one on the right) or against a marker post which is fitted to some machines at a measured distance from the reproducing head. The tape is then removed from the head and placed in a special guide, usually made of aluminium, which has a lip slightly narrower than the tape. The guide has a diagonal slit at either 45° or 60° to the tape. The mark on the tape is lined up with the slit, or with a mark on the guide at the same distance from the slit as the market post is from the repro head, and cut by running a razor blade along the slit. The angle of cut is chosen to suit the programme material. The more diagonal cut creates a smoother effect but is less precise. The 60° is mainly for removing clicks.

Joining the tape

The recording tape or leader tape to which the first piece is to be joined is laid in the guide and cut in a similar manner. The reason for the diagonal cut is to create a smooth transition and prevent a click occurring due to a sudden change of signal level.

The two tapes are then butted together and about three centimetres of special adhesive tape is stuck over the join on the back of the tape. Care should be taken to put the adhesive tape on straight and to touch the sticky side as little as possible with the fingers.

Leader tapes

One convention is to use white leader tape (about one metre or three feet) at the start and to write the title on it, then short lengths (about 15 cm or 6 in) of yellow leader between separate inserts, and red leader tape at the end. It is a good idea to get into a routine of marking the tape for splicing in a particular manner, e.g. always along the bottom of the tape. This avoids confusion and the risk of joining the tape on backwards when the splice is made. If it is known that a recording is to contain a number of short inserts but there will not be time to make joints in the tape (e.g. for newsreel) time can be saved by splicing the tape beforehand with rather more tape between the leaders than will be required. The sequences are then recorded immediately after the leader in each case and this forms a marker for each insert.

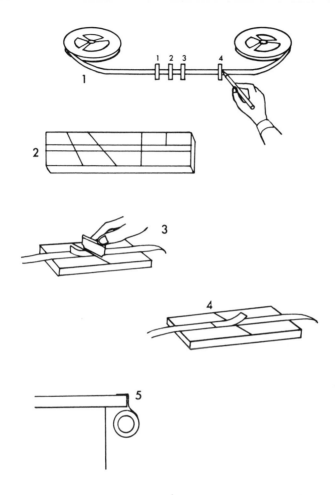

1. Professional tape-machines have three heads: the erase (wipe) head on the left, record head in the centre and playback head on the right. The editing point is found by 'rocking' the spools backward and forward. It is then marked with a yellow wax pencil either opposite the replay head gap or against a special 'marker post'.

2. An editing block with provision for cutting at 60°, 45° and 90°. When the block is used in association with a tape recorder with a marker post, a mark is incorporated on the block at the same distance from the slit as the marker post is from the repro head.

3. The tape is cut with a non-magnetic blade together with the piece of tape to be joined.

4. About one inch of joining tape is laid carefully across the join after removing the top cut off piece.

5. Operators usually stick the roll of the tape to the side of the machine for quick action. The area touching the table is cut off before use.

Multi-track Recording

Efficient post-production work and music recording require the use of a multi-track recorder. There are analog multi-track machines that record 4 or 8 tracks on 25.4 mm (1 in) tape or 16, 24 or 36 tracks on 50.8 mm (2 in) tape. There are also digital multi-track machines (page 218) and digital memory recorders using magnetic discs.

The purpose of multi-tracking
There are three basic reasons for multi-tracking:
1. To enable re-mixing and various forms of manipulation to be accomplished at a later date without wasting recording time.
2. To enable artists to play in accompaniment to their own performance, to increase the apparent number of artists on the final recording, allowing for repeat takes of each addition without spoiling the previous ones and without unduly increasing the noise level.
3. To enable recording machines to be synchronised with other tape recorders, videotape recorders or film machines, using one track to carry a synchronising signal or time code. In this way sound tracks for video recordings or films can be built up, adding dialogue, effects, music etc. on the various tracks, all synchronised with the picture. They can then eventually be mixed down to form a single synchronised sound track.

Multi-track recording procedure
Making a multi-track master recording involves multi-microphone technique. Each section of the orchestra or instrumentalist in a group is given a separate microphone, sometimes more than one to each player (e.g. a drum kit can require up to 10 microphones for some types of balance). The microphone outputs are processed in a multi-channel sound console which provides the facility to monitor individual channels (by selection) or produce a 'dummy' mix in which they are mixed into the monitoring system while remaining separated for the recording tracks. Visual monitoring is provided for all channels so that the volume on each of the tracks can be kept within the parameters of the recording system.

At a later stage the various tracks are 'mixed down' to mono, stereo or quadraphonic form. This process can take a considerable time if much editing is involved or if complicated treatment (e.g. various forms of artificial reverberation and frequency response shaping) is required for the individual tracks. It can all be accomplished, however, without retaining the artists.

Multi-track recorder mechanisms
Tape recorders handling wide tapes require powerful motors, well engineered mechanical systems and electronic tape tension sensors because of the large transfer of weight as the tape spools from one reel to the other.

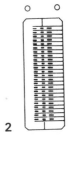

1. A typical 24-channel multitrack recorder. Note the VU meters for each channel. In order that the tracks can be recorded at different times but in sync it is necessary to employ a complicated system of head switching. Individual erase heads are provided for each track.
Although separate record and replay heads are provided, arrangements have to be made to use the record head to playback previously recorded tracks while further tracks are being recorded, to enable the artist to synchronise with the previous recording. (The repro head does not synchronise with the record head.)

2. Typical multitrack record head. Great care must be taken in setting the azimuth, otherwise serious phase differences can exist between the most separated tracks. When selecting tracks for recording it is advisable to place any sources that are likely to have overlapping sound (e.g. two microphones in close proximity) on adjacent tracks. Otherwise phase differences could cause undesirable effects.

203

Dynamic Noise Reduction: Dolby A System

Modern recording techniques usually involve the re-recording (dubbing) of programme material from one medium to another, in some cases many times over. This can be for the purpose of 'reducing' a multi-track recording to twin-track stereo, for editing or for transfer to disc etc. Unless digital recording is used, each recording will add its quota of noise to the final result.

Companding

In order to improve the signal-to-noise ratio a system of 'companding' can be used, i.e. the volume range of the signal is compressed to raise the average level before recording and then expanded on reproduction, depressing the quieter elements of the programme (and with it the noise).

Unfortunately there are two main problems in designing a companding system that does not cause more trouble than it cures. One is the difficulty of making the compression and expansion systems 'track' perfectly over the whole frequency and volume range. The other is preventing the background noise level going up and down with the signal, which only makes it more obvious. The *Dolby A system* overcomes this problem in the following ways.

The signal is provided with two parallel paths, one via a linear amplifier and the other via a differential network, the output of which is added to the 'straight through' signal when recording and subtracted when reproducing. The differential network divides the frequency spectrum into four bands and treats each one separately so that the action is applied only where it is needed. With music, for instance, most of the sound energy tends to be concentrated in the lower middle band, whereas tape hiss is largely high and too far removed in frequency for 'masking' to take effect. This is the effect whereby the ear tends to be deaf to sounds in the presence of, and for a short time after, hearing louder sounds of similar frequency. The Dolby system exploits this effect by restricting the action to narrow frequency bands.

The system is adjusted so that with a low level input (below −40 dB) the output is 10 dB higher than the input for frequencies up to about 5 kHz, above which it increases progressively to 15 dB gain at 15 kHz. As the input signal rises above +40 dB, it is progressively less affected. At high levels (where the possible ill effects of companding are most obvious), the compander has least effect and is virtually bypassed. The Dolby A system is normally used only as an intermediary process, the final product having a straightforward response. This is in contrast to the Dolby B process, where the recording is sold with the modified characteristic incorporated on the assumption that the complementary characteristic will be applied in the reproducing apparatus.

204

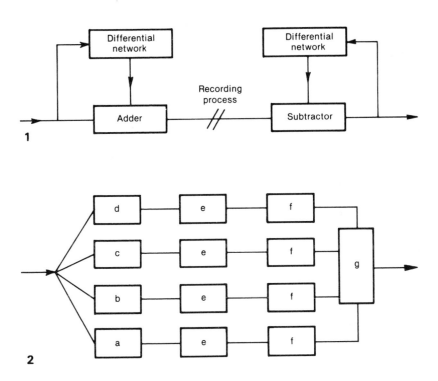

1. Companding the recording process. The output of the differential network is added for recording and then subtracted when replaying.

2. Schematic detail of differential network: (a) 80 Hz low-pass filter to deal with hum and rumble; (b) 80–3000 Hz band-pass filter to deal with mid-band noise and print-through; (c) 3000 Hz high-pass filter to suppress hiss and modulation noise; (d) 9000 Hz high-pass filter to suppress hiss and modulation noise; (e) linear limiters; (f) non-linear limiters; (g) adder.

Dolby B Noise Reduction System

A simpler system than the Dolby A process is the Dolby B system which is used mainly in the domestic tape cassette market to reduce the high background level of tape hiss which is inherent in this type of recording due to the low tape speed and narrow track. With Dolby B the tapes are recorded and marketed with the Dolby characteristic on the assumption that they will be reproduced by equipment with the complementary characteristic.

As in the Dolby A system, there is a main programme path and a side chain. The side chain incorporates a compressor which is preceded by a high-pass variable filter covering frequencies from about 500 Hz upwards. In the record mode, signals below a given threshold level are boosted by the compressor and added to the side chain. This increase in level is applied progressively by the variable filter from 500 Hz up to 10 dB at 10 kHz.

Thus low-level high-frequency signals are recorded up to 10 dB higher than the original. An overshoot suppressor (diode clipper) is provided, following the compressor, to prevent high-level transients which are faster than the time constant of the compressor from being added to the output. Any resultant distortion is masked by the high-level signal and tends to cancel on replay.

The Dolby B decoder, which is fitted to most high-quality cassette machines, is identical to the coder except that the side chain is fed with a phase-inverted signal so that the output of the compressor is added to the main chain in antiphase and is effectively subtracted from it, thereby producing an exact complement of the coding process.

Noise reduction

As the low-level high-frequency response is reduced on playback so is the tape hiss and system noise that has been introduced by the recording, giving an improvement in signal-to-noise ratio of up to 10 dB.

The difference between the Dolby system and the simple application of pre-emphasis to the recording and de-emphasis on playback (which is applied anyway) is that the Dolby B characteristic affects only low-level signals.

Compatibility

Dolby B encoded material can be played on equipment without the Dolby function if the high-frequency response is reduced to compensate for the Dolby characteristic, but this does result in a lack of high frequency in the louder passages of the music.

206

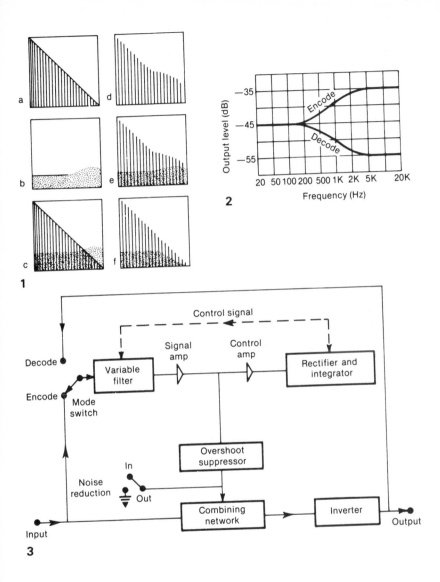

1. (a) A representation of the distribution of energy in average programme material. The energy tends to decrease as frequency increases. (b) Tape noise. (c) In normal reproduction low-level high-frequency signals are masked by tape hiss. (d) Dolby B encoded signal with compression applied to boost low-level high-frequency signals. (e) When the tape noise is added the boosted high-frequency signal is still above the noise level. (f) Decoding reduces the low-level high-frequency response to normal, and with it the noise.

2. Complementary response curves of a Dolby processor with low-level input.

3. Block schematic diagram of a Dolby B encoder/decoder.

An extension of the Dolby B principle can produce even more effective noise suppression.

Dolby C Noise Reduction System

Dolby B (page 206) is a noise-reduction process in which 'sliding-band' compression is applied in the making of a recording, on the assumption that complementary expansion will be used in reproduction. It provides up to 10 dB of noise reduction.

Dolby C is an extension of the Dolby B principle. It offers up to 20 dB of noise reduction with improved protection from recording distortions and unwanted side effects. It employs two compressors in series in the recording process and two complementary expanders in reproduction, the first stage operating at levels comparable with B-type systems (which can make use of similar circuitry) and the second sensitive to signals 20 dB lower in level. The result of adding the two companders is to increase the noise reduction to 20 dB in the critical 2 to 10 kHz area. Whereas the type B system begins to take effect in the 300 Hz region and increases its action to a maximum of 10 dB at 4 kHz and above, Dolby C begins to take effect in the 100 Hz region, providing 15 dB of noise reduction around 400 Hz; it thereby reduces the 'midband modulation effect' in which middle frequencies become modulated by high-level high-frequency signals, owing to a mistrack between the encoded and decoded signals.

Spectral skewing

The problem of midband modulation and other unpleasant effects, due to the unpredictable response of cassette recorders at the extremes of their frequency range, is overcome by a process called 'spectral skewing'. This consists essentially of a steep roll-off in the input to the high-level stage of the compressor to make it insensitive to signals above 10 kHz. A complementary boost above 10 kHz is applied at the output of the final expander.

Antisaturation

To reduce the risk of overload and tape saturation at high frequencies, with consequent risk of compander mistracking, a shelving network is placed in the circuit which affects only the high-level signals. At low levels most of the signal passes through the side chain. A complementary network is provided in the expansion mode to maintain a flat response.

Owing to the two-stage design of the Dolby C system it is relatively simple to design equipment with the facility to switch between Dolby B and C.

208

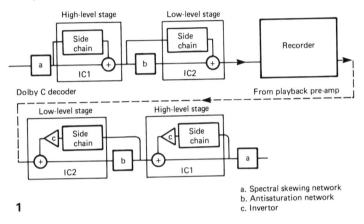

a. Spectral skewing network
b. Antisaturation network
c. Invertor

1

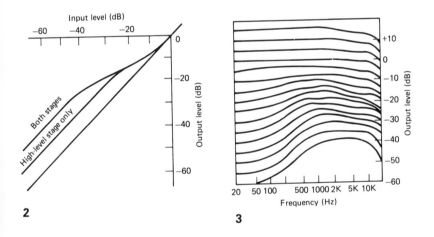

2

3

1. Block schematic diagram of a Dolby C encoder/decoder.

2. Input–output transfer characteristic of C-type compressor at 1 kHz, showing the effect of high-level stage on its own, and both stages in series.

3. Dolby C compression characteristics.

DBX and DNL noise suppression systems rely on the premise that high frequency energy tends to increase disproportionately with a general increase in volume.

The DBX and DNL Noise Suppression Systems

DBX noise suppression, like Dolby A, is a complementary system.

DBX

The DBX system employs a wide range (2:1) compression ratio to encode and decode. Problems of tracking the compression and expansion are eased by the straightforward ratio and the fact that the level sensing is done on an RMS basis, i.e. controlled by the total power of the signal, regardless of the phase relationships of the various components.

The DBX system exploits the fact that, in most programme material, the bulk of the power is in the low frequencies, high power in the high frequencies occurring only when the general volume is large.

Pre-emphasised compression

The signal fed to the compressor is heavily pre-emphasised to increase the overall power on recording. It is similarly de-emphasised to restore it to normal (while at the same time reducing the high-frequency noise) when it is decoded. To prevent the recording being overloaded by any powerful pre-emphasised high-frequency signals, a similar pre-emphasis is applied to the side chain of the compressor so that, for high levels, the high-frequency recording level decreases as the frequency increases and increases as the high-frequency level decreases.

Masking effect

DBX also takes into account the masking effect of our ears, which makes us relatively insensitive to noise (in this case high-frequency noise) in the presence of loud sound of a similar frequency. The use of DBX can result in an improvement in high-frequency signal-to-noise ratio of up to 30 dB.

Philips DNL system

The Philips DNL system operates on replay only, and therefore must reduce the fidelity of the recording to some extent. As most musical instruments have fundamental frequencies below 4.5 kHz and when played softly produce few harmonics, it is considered that a high-pass filter operating above this frequency has little effect on quiet material.

The system is basically a dynamic treble cut filter operating above 4.5 kHz on low amplitude signals. When the signal exceeds the predetermined level, the filter is by-passed thus leaving the harmonics unaffected.

The system is claimed to give an improvement in signal-to-noise level of 10 dB at 6 kHz, 20 dB at 10 kHz.

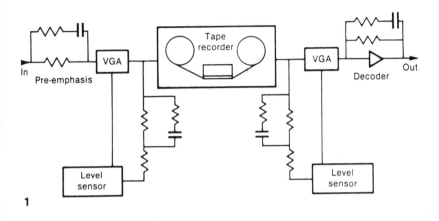

1

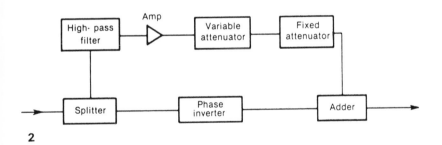

2

1. Block diagram of the DBX noise reduction system. The variable gain amplifiers have a compression/expansion ratio of 2:1. Due to pre-emphasis, the increase is applied progressively to the high frequencies.

2. Block diagram of the Philips Dynamic Noise Limiter. This works on replay only and reduces the high-frequency response when the high-frequency component of the signal is low.

Sound and Electronic Editing

The cutting of tape is being superseded by electronic editing. For this purpose a VTR machine fitted with an electronic editor is required. This is a device that controls the switching logic to make allowance for the distance between the erase and record heads so that the machine can be switched instantly from replay to record, or vice versa, without losing signal continuity.

Studio editing

The studio editing proc;ess is as follows. At the 'out' cut point at the end of a studio sequence the director presses a button which records a cue pulse on the tape and the machine is stopped. The next sequence is then set up and the machine wound back about 20 seconds. The VTR machine is then restarted and, if required, the sound is fed back to the studio to help the actors to pick up their cues. About 1.3 seconds before the editing point a short burst of tone known as the 'advance cue' is relayed by a loudspeaker in the production control room. It can be heard by everybody listening to talkback as a warning that the cue is coming. When the cue pulse is reached the studio monitors automatically switch in vision and sound from the VTR to the studio output. This enables the director to judge the effectiveness of the edit. If it is not correct the cue pulse can be re-recorded or, if the error is slight, the VT switch/pulse time relationship can be altered by up to 12 or 18 frames (according to machine type). When the correct edit point is established the process is repeated in the record mode and this time the VT recorder switches to record when the cue pulse occurs.

Insert mode

The above operation can be carried out in vision or sound separately, so that it would be possible to insert a new piece of sound track or change the vision while allowing thet original sound to remain.

Dub editing

Most electronic editing concerns the assembly of pre-recorded tapes by dubbing from one machine to another. It is possible to make minor adjustments to the sound in the process, but for large alterations, e.g. the addition of incidental music, the sound track is dubbed separately after the vision has been edited. This could involve the use of a multi-track audio recorder, on one track of which an address code is recorded to synchronise with an identical code on the cue track of the VTR. A copy can also be made on a less expensive helical-scan VTR (also address-code synchronised) and all operations conducted, using this as a vision reference, before finally dubbing the sound back to the broadcast VTR.

212

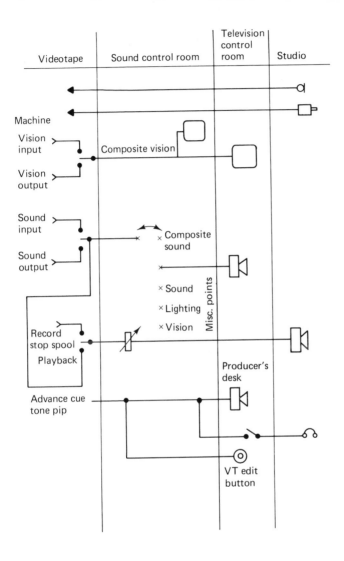

Interconnections for studio electronic editing. The switches (which are in fact a ganged relay) are operated by the edit cue recorded on the tape. The up position is the record mode, the down position the playback mode. The monitors show either the input to the VT machine or its output according to the position of the switch. The studio loudspeaker is effective only on replay. The advance cue pip is fed to the producer's desk loudspeaker and also, if requested, to all people listening to talkback.

Digital Audio

Processing audio by digital means offers enormous advantages over analog systems in terms of signal-to-noise ratio, robustness, freedom from degradation in multicopying, transmission and multiplexing etc.

In a digital system the original analog signal is converted into a code comprising a series of pulses, representing numbers in binary notation which correspond to the amplitude of the waveform at each moment in time.

Analog to digital conversion (ADC)

The process of converting from analog to digital format consists of:

1. *Sampling* Samples are taken of the instantaneous value of the analog waveform. They are taken at a fixed rate which must be at least twice the highest frequency that it is intended to convert. This because of the *Nyquist limit.* If the converter is presented with an analog input of more than half the sampling frequency a form of heterodyning occurs, producing a spurious frequency called *aliasing* (see page 216). The standard sampling rate for domestic digital equipment (such as CD players) is 44.1 kHz and for professional equipment 48 kHz. Each sample is held in a store while it is measured and converted into a code.

2. *Quantising* The process of measuring each sample involves comparing it against a scale consisting of a number of discrete values called *quantising levels*, each of which is assigned a reference number to represent its value. As the analog signal can have values which vary continuously over the whole range of measurement the actual level and the quantising level will seldom coincide exactly. The greater the number of quantising levels the more accurate will be the measurement. Differences between the sample levels and the quantising levels can be heard as noise in the reproduced sound. As the error is proportionately greater with fewer quantising levels, the noise tends to be more objectionable at low levels.

3. *Coding* The quantised levels are converted into a code, usually binary, which represents them as a group of digits composed of 'on' or 'off' signals. A binary code with 10 digits will provide 1024 (i.e. 2^{10}) quantising levels with a signal-to-noise level of 60 dB. The 16 bit code as standard for CDs produces 65 536 levels with a signal-to-quantising noise level in excess of 90 dB.

Digital to analog conversion (DAC)

In DAC the digital code is converted to a series of levels corresponding to the original samples, presented at the original sampling rate. The analog output is passed through a low-pass filter to inhibit the harmonic images of the sampling frequency (see page 216).

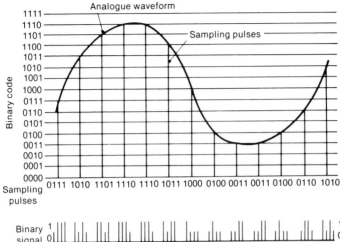

1

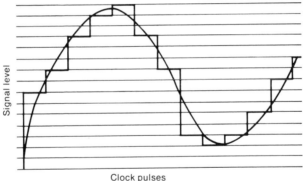

2

1. An example of sampling and coding of an analog signal. For simplicity only a 4 bit code is shown, which would allow only 16 levels of measurement (quantising levels). This would result in a high degree of error and thereby a high level of quantising noise. This takes the form of 'white noise' (i.e. noise with an equal spectral distribution) on large amplitude signals and 'granular distortion' (resembling non-linear distortion) at low levels where only a few quantising levels are in use. For high-quality sound reproduction, codes with about 13 bits are used. A 13 bit code would provide 8192 quantising levels and a signal-to-noise ratio of 78 dB. Some systems use logarithmic instead of linear sampling so that there are more quantising steps at low level where it is more critical.

2. Digital to analog conversion. The digitised signal is converted to a series of levels corresponding to the original samples presented at intervals ('clocked') corresponding to the pulse rate of the original sampling process. This produces a jagged waveform which, after passing through the low-pass filter, resembles the original analog wave.

Oversampling and Error Correction

Nyquist limit

In discussing analog to digital conversion (page 214) it was pointed out that the sampling rate must be at least twice the highest frequency it is required to record, because of the *Nyquist limit.* If too high a frequency is input into the system it will create a beat (difference) frequency between the input and the sampling frequency. This produces a sum and difference effect which is rather like the way spoked wheels appear to be turning backwards on TV when the spoke frequency of the wheel is slightly different from the sampling rate of the camera. This effect is *aliasing.* It produces spurious noises in the sound. The effect will occur at each multiple (harmonic) of the sampling rate, so to prevent aliasing a very steep 'brick wall' low-pass filter, with a cut-off frequency of half the sampling rate, would have to be inserted in the analog output. Unfortunately such a steep filter would introduce serious phase nonlinearity, which could negate much of the advantage of digital recording.

Oversampling

The problems of aliasing can be overcome by increasing the sampling rate to much greater than that required to satisfy Nyquist. This would push the first harmonic much further up the scale so that a much more gentle low-pass filter could be used. For example, *four times oversampling* places the first harmonic in the spectral position where the fourth would have been, so that a filter of 12 dB per octave would be adequate.

Oversampling in CDs is achieved by digital filters, which effectively increase the sampling rate.

Error correction

A major advantage of digital over analog recording systems is the ability to correct or conceal errors, such as can be caused by tape drop-out or dirt on a CD. The error is identified by checking the signal for parity.

Parity checking consists of adding a parity bit to each binary number to make the number of 1s even (or odd). If an inversion has occurred the number of 1s will not be even (or odd). The parity check will detect this and will either repeat the previous sample or give a value between the previous and following value. This is known as *error concealment.*

Error correction as applied to CDs actually reconstitutes the missing data. The data is sent out in blocks with parity checks interleaved in such a way that a burst of errors (as in dropout) would be spread along the data stream.

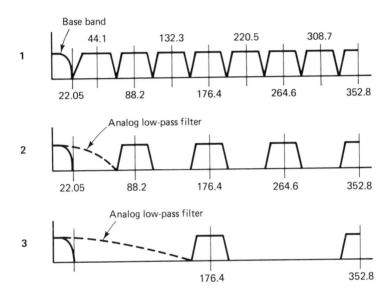

Digital oversampling filter. 1, the unfiltered response of a DA converter contains an infinite series of images of the original signal audio-band spectrum. 2, digital filtering with two times oversampling. 3, four times oversampling.

Oversampling suppresses the images immediately above the audio band and the remaining images can be filtered out with a relatively less steep filter.

A typical four times oversampling filter could consist of a shift register of 24 delay elements, each delaying a 16 bit sample for one sample period during which it is multiplied four times, with a different coefficient used for each multiplication. The four sets of coefficients are applied to the samples in turn, producing four output values per sample. The 24 multiplication products are summed four times during each period and the product becomes the output from the filter, effectively increasing the sampling frequency by a factor of four. The output of the digital filter is converted to analog and the remaining band around 176.4 kHz is completely suppressed by an analog filter which need only have a 12 dB per octave slope. This can be achieved without phase distortion.

217

Digital Audio Magnetic Recorders

Digital audio magnetic recording offers a flat frequency response throughout the audio range, with complete freedom from tape hiss, wow and flutter and modulation noise. As, in digital form, only the two binary states (represented on the tape by S–N and N–S) have to be recorded, a very high packing density can be achieved so that tape speeds do not have to be excessive.

Digital Audio Static Head (DASH) recorders

These are similar in appearance to analog tape recorders. The tape transport is fundamentally similar but more care is necessary over speed control and tape tension because the digital recording tape is thinner and smoother in order to provide a good head contact for HF response. The smoothness of the tape tends to trap air during spooling, so digital recorders do not spool as fast as analog ones. The capstan is controlled at constant speed when recording unrecorded tape, but on repro is controlled by a reference signal off the tape or from a synchroniser, possibly using time code, by which several machines can be run in synchronism. The control track also provides information for autolocation and specifies the type of format and sampling rate in use. Three sampling rates are available: 32 kHz, 44.1 kHz and 48 kHz.

Error correction and time base correction are used. The data code is cross-interleaved with delays so that dropouts and other tape errors are spread along the tape and can be interpolated between samples to provide correction.

Punch-in and out editing

The process of punching-in new material to replace a section of recording, or assemble editing, is complicated by the data interleaving, because each moment in time is represented by a spread of data along the tape. So a crossfader is used, which deinterleaves the replay signal just before the edit point and reinterleaves it afterwards. The speed of the crossfade can be chosen to suit the programme material. The edit can be rehearsed, with only the monitoring changing over, before putting the machine into record.

Cut and splice editing

This is possible with digital tape. The tape must be cut at right angles. Splicing upsets the code sequence, but this occurs at different points in the odd and even sequence, so that by interpolation it is possible to obtain the end of the old recording and beginning of the new one simultaneously, enabling the crossfader to make a rapid digital crossfade while both are available.

218

Types of DASH recorder

There are several versions of the DASH format, including stereo and multitrack machines.

DASH-F is the fast speed version which runs at 30 in/s, providing 24 audio tracks on half-inch tape.

DASH-M is medium speed (15 in/s) with each audio channel spread over two tracks, and is double recorded, so that the eight tracks provide two audio channels double-recorded on quarter-inch tape.

DASH-S is the slow-speed version. Each audio channel is spread over four tracks with the tape running at 7.5 in/s, providing two audio tracks on quarter-inch tape.

Twin DASH means that the data for each audio channel are recorded twice, which makes the recording more able to cope with cut-splice editing.

DASH II is the double density version, which uses thin-film heads to achieve 48 digital tracks on half-inch tape or sixteen tracks on quarter-inch tape. The tracks are so aligned that DASH I recorded tapes can be played on a DASH II machine.

Examples of machines in current use

SONY PCM 3324 is codified as DASH-FIH:

*F*ast format, one channel per track. Single density (I).

*H*alf-inch tape. 24 tape tracks and 24 audio channels.

PCM 3202 is a twin DASH-MIQ machine:

*M*edium format, two tracks per channel. Single density.

*Q*uarter-inch tape. Eight tracks, two audio channels double-recorded.

PCM 3102 is a DASH-SIQ machine:

*S*low format, four tracks per channel. Single density.

*Q*uarter-inch tape, eight tape tracks, two audio channels.

The machines have built-in time code generator/readers and the ability to lock to incoming time codes such as SMPTE/EBU.

Recording Audio on Video Recorders

Magnetic tape recording in analog form is to some extent a compromise between tape speed, HF response and head-gap size (which also affects bass response) and is prone to tape hiss modulation noise, wow and flutter. All these problems can be overcome by digital recording.

Digital recording

Unfortunately the bandwidth required for a digital recording (derived by multiplying the sampling frequency by the number of bits generated in each sample) is outside the range of a normal audio recorder, although specialised recorders are available (page 218) which achieve the necessary packing density. An alternative method of obtaining the necessary bandwidth is to record audio in digital form on a video recorder, making use of the high head/tape speed afforded by the rotating head.

PCM adaptors

PCM adaptors are available which convert the audio into a pseudo-video signal, complete with sync pulses to lock to the format of the VTR. Black level represents binary 0 and about 50% of peak white represents binary 1. To cover the interruption caused by the sync pulses the digital samples are fed into a memory, read out at a rate which is higher than the sampling rate and then serialised in the correct order.

It is advisable to use a video recorder with a 'PCM' or 'Digital' switch which disables its drop-out compensator, because the PCM adaptor produces its own error correction and the addition of the two could cause spurious noises.

Editing

Editing a videotape is not a simple matter of cutting and splicing as with analog tapes. The proper relationship between the sync pulses of the recordings has to be maintained, and anyway splicing helical scan recordings is impracticable because of the long slanting nature of the tracks.

Video and digital recordings can be edited by dubbing from one machine to another. This requires a sophisticated electronic synchronising system which involves recording a digital time code on the cue or address track of both machines. The machines can then be run locked together at the edit point, maintaining the proper sync pulse relationship throughout. The new sequence can be butt-joined or cross-faded to the previously recorded one and levels adjusted in the process. As the recording medium is digital it suffers no degradation in the dubbing process.

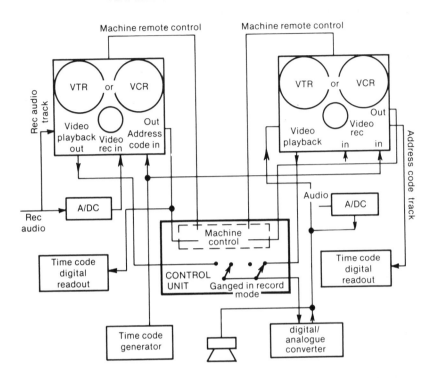

PLAY-IN MACHINE   RECORD ASSEMBLY MACHINE

Schematic representation of two-machine time-code editing for digital sound. The control unit controls the mechanical functions of the playback machine and the record machine. A time code recorded on the address code track of each tape gives visual indication of their position. When both tape machines are locked to the synchroniser in the control unit they can be run in replay, record or spool in perfect sync. The synchroniser can also be used to locate a particular position on the tape.

Before editing, the machines automatically set to a position a few seconds ahead of the editing point. Editing is achieved by running the two machines together and playing in the selected sequence at exactly the required edit point on the master tape set by the time code. The edit can be checked by running through the transition in the 'rehearse' mode, when only the monitoring changes over at the edit point. If this is not exactly right, the edit point can be moved (by offsetting the electronic lock-up) and only when it is considered perfect is the device put into the 'record' mode. When the edit point is reached, the record machine goes into record and the edit is completed.

Cue and address code

Using a standard VCR the audio signal can also be copied to the normal (longitudinal) audio track to help in identifying the sequences; the address code, on another longitudinal track, can provide a digital display of tape position.

Sound on a Videotape Recorder

For many years TV sound has suffered from the indifferent standard of audio recording on videotape. This has been due to the relatively slow tape speeds (by professional standards), narrow tracks, and speed controlled by video tracking. All these problems can be overcome by digital audio recording.

C-format recorders

The C-format is a helical-scan rotary-head machine. Early NTSC C-format machines have three longitudinal analog audio tracks, two on one edge of the tape and one on the other. In PAL/SECAM machines an additional audio track is available if the vertical interval of the video is not recorded.

Digital C-format machines

In the digital C-format machine a digital recording is laid down in the area of tape available for a fourth audio track (or syncs) by digital heads in the rotating drum. The tracks occupy only 22° of the drum rotation, so the digital signal has to be time-compressed to fit all the audio samples in a field period into the available space. Three digital tracks can be fitted into the space of one analog sync track. They can be narrow because signal-to-noise ratio is not a problem. The resulting bandwidth allows two digital audio tracks to be recorded, together with the necessary correction signals. The C-format uses three video heads in the drum for erase, record and confidence monitoring. Three heads are also used for the digital audio for advanced playback, record and confidence/replay. The system can only work at standard speed, otherwise the longitudinal track must be used.

D-1 and D-2 cassette recorders

There are two different formats for digital casette machines: colour-difference (D-1) and composite-digital (D-2). The colour-difference format can use 16 µm or 13 µm tape and the composite digital only high-coercivity 13 µm tape. It uses offset azimuth recording, which enables a closer track spacing, resulting in a considerably longer playing time. There are three sizes of cassette. The small size allows 11 minutes on 16 µm tape or 13 minutes with the 13 µm tape on the D-1 format, or 32 minutes on the composite format. The medium size cassette provides 34, 41 and 99 minutes respectively, and the large cassette 76, 94 and 221 minutes.

Both formats convey four audio channels in digital form. In both recorders the same heads are used for video and audio, but in the D-1 format the video signal is split into two segments on the track with the blocks of audio data in the centre, whereas the composite has the audio at each end of the tracks. Split audio/video and assemble editing is possible.

222

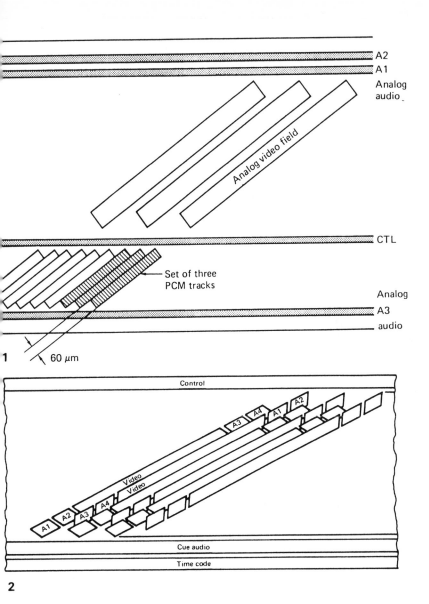

A2
A1

Analog
audio

Analog video field

CTL

Set of three
PCM tracks

Analog
A3
audio

1 60 μm

Control

A3 A4 A1 A2

Video
Video

A1 A2 A3 A4

Cue audio

Time code

2

1. Track arrangement in the digital audio version of C-format. The PCM audio tracks are recorded in the area used for syncs or fourth audio channel in the previous version. The azimuths of the digital tracks are offset so that when played on an analog audio machine they cancel.

2. Track arrangement for D-2 cassette machines. The twin pairs of heads produce a staggered pattern with the audio blocks at each end. Offset azimuth recording enables close packing of tracks due to cancellation of crosstalk.

In both D-1 and D-2 the audio signals are recorded twice in different positions along the tape, to give immunity to head clogs and scratches.

Video-8 Audio Recording

The video-8 format is a helical scan rotary head recorder. Two longitudinal audio tracks are provided, but as they are narrow and as the tape speed is only 20.05 or even 10.06 mm/s they are not suitable for high quality recording.

FM carrier

An FM carrier system is provided which supports a single audio channel in the video waveform. This is good quality but monophonic and cannot be divorced from the video for editing purposes.

PCM recording

Video-8 can also offer PCM recording in stereo. This is achieved by extending the half-cylindrical wrap of the tape around the drum by a further 31 degrees, in which space the time-compressed audio signal is recorded. The drum contains two heads so that while one is playing the last 31 degrees of a video field the other will be playing the audio at the beginning of the next. To accommodate the extension of the diagonal tracks the tape is made 1.25 mm wider.

As the video and audio are coded separately they can be recorded separately, possibly at different times.

8 mm audio-only recording

The video-8 format can be used as an audio-only recorder, with the possibility of a long playing time. The ability to time-compress the audio signal into 31 degrees of head rotation opens up the possibility to record six audio segments in one track. Only one segment in each track is recorded as the tape passes, but when it reaches the end of its travel it reverses and records the next segment in the opposite direction, and continues to record in alternate directions until all six segments are filled.

Due to the track-following mechanism it is not possible to play back more than one segment at a time, as there is no fixed relationship between them, so the system cannot be used as a multi-track recorder.

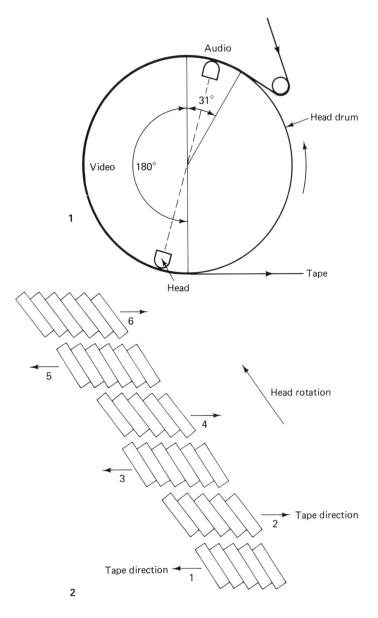

1. Video-8 head wrap. Video occupies 180° of drum rotation, audio occupies 31°. Tape is made 1.25 mm wider to accommodate extra track length.

2. 8 mm audio-only recording. Six recordings are made in alternate directions, the tape reversing between each. The recordings do not line up with each other, so only one can be used at a time.

Audio signals, converted into digital form, can be recorded as a series of pits arranged in spiral form and read with a laser beam.

Compact Discs

The compact disc is a very convenient method of reproducing high-quality pre-recorded sound. It is robust to handle and is capable of providing a flat frequency response with signal-to-noise ratio and channel separation in excess of 90 dB.

The recording consists essentially of a digital code in the form of a series of pits 0.5 μm wide and of eight different lengths varying between 0.833 μm and 3.56 μm in the surface of a 12 cm plastic disc. The surface of the disc is plated with a reflective layer and a plastic coating, which also carries the label, is applied on top. A laser beam is used to read the pits through the plastic disc from underneath (where they appear as bumps). The laser beam is focused on the bumps and is reflected back along the optical path where it is deflected by a half-silvered mirror to a photo diode to provide the data output. As the data can contain a number of adjacent identical bits it is necessary to employ a modulation system to discriminate between successive symbols and also to reduce the DC component in the output. Otherwise the data signals would interfere with the focus and tracking mechanisms, which rely on relatively slow changes in output for correction.

Tracking
To record up to 74 minutes of programme as well as a run-in and run-out section on a 12 cm disc involves the storage of an enormous amount of data (over 6 billion bits). So the separation between the helical train of pits is only 1.66 μm; to achieve maximum packing density the spacing between the bits is maintained constant by continuously varying the rotational speed of the disc to achieve constant linear velocity. This is done by locking the disc rotation servo to the data rate from the disc, as the laser beam tracks the pits from the inside working outwards.

Tracking and focus of the laser beam have to be extremely accurate. This is achieved by servos controlled by the reflected beam. The depth at which the reflective surface is contained within the plastic disc ensures that blemishes on the surface are sufficiently out of focus not to affect the reproduction.

In addition to the programme recording, eight separate sub-codes (nominated P, Q, R, S, T, U, V and W) are multiplexed. The P channel is used for the pause signal and muting between tracks and at the end before the lead-out. The Q channel is used to switch the required pre-emphasis circuits. These codes enable circuitry to select start points on the disc.

Professional CD machines
With domestic CDs there is usually a small time lag between the start point on the disc and the beginning of the audio. Professional CDs provide the ability to find the exact point required by rocking back and forth with a search dial. It can then be stored in a memory for instant cueing.

226

Too close

$(1+3)-(2+4) > 0$

Too far

$(1+3)-(2+4) < 0$

Correct

$(1+3)-(2+4) = 0$

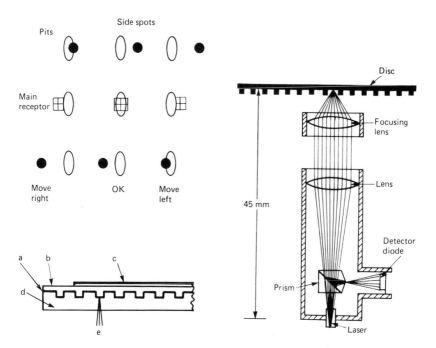

Side spots

Pits

Main receptor

Move right

OK

Move left

Disc

Focusing lens

Lens

45 mm

Detector diode

Prism

Laser

a b c

d

e

1. *Focusing mechanism.* The focus of the laser beam spot is maintained by an optical system which throws a light pattern in the reflected beam, the pattern changing with distance. The photo diode receptor is split into four segments arranged as in the diagram. The output of detector 1 + 3 is subtracted from 2 + 4 and the difference signal is fed to a servo mechanism which moves the optical system to maintain it at zero (initial focus having been established by a ramp current which moves the objective lens over its entire travel).

2. *Tracking mechanism.* The combined output of the four detectors provides the digital readout from the disc and is also combined with two side-spot detectors to guide the servo that moves the optical system laterally to track the pits in the disc. The side spots are produced by a diffraction grating adjacent to the laser diode.

3. *Section through compact disc.* (a) Reflective layer. (b) Protective film. (c) Label. (d) Plastic carrier. (e) Laser beam.

4. *Optical pick-up of the Philips compact disc.* The laser light, focused on the alluminised reflective coating at the depth of the pit, is reflected back along its path to be diverted by the prism into the detector. Where there is no pit the beam is defocused and much less light is returned.

Digital techniques can be applied to audio recording on cassettes and to sophisticated cueing methods.

Digital Audio Magnetic Recording

The challenge to produce a recorder with the quality potential of the compact disc and the ability to record has led to the development of the R-DAT recorder. The requirement for flexible cueing has led to the Audiofile.

The R-DAT cassette recorder
R-DAT stands for rotary-head audio taperecorder, which records audio in digital form on a cassette measuring only 74 × 53 × 10 mm (2.9 × 2.1 × 0.4 inch), much smaller than a standard audio cassette and similar in construction to a video cassette. The heads are in a drum which rotates as in a video recorder, with a wrap angle of only 90° to facilitate lace-up.

The 7 μm thick metallic tape runs at 8.1 mm/s and can accommodate up to three hours of stereo with 16 bit quantising and 48 kHz sampling. Tracking and index data are included with the sub-code data at about 275 kbit/second. Two low-grade analogue tracks are also available. The head drum rotates at 2000 rev/min and the tape, which is very light and delicate, is carefully controlled at every point by servos. The headpressure is only 1/10 gram, and the head remains in contact during spooling so that the control signals can be read continuously.

Domestic R-DAT machines record only at the 48 kHz sampling rate and cannot therefore record directly from CDs without going through the analogue process. As a further inhibition to dubbing, each R-DAT machine records its serial number in the sub-code data.

The facility to record time-code data in the sub-code enables professional versions of the machine to synchronise with video equipment and film cameras etc. This makes them eminently suitable for synchronous recording 'in the field', so they may eventually supersede the SEPMAG recorder.

The Audiofile
Audiofile is a method of recording 16 bit digital audio on hard 'Winchester' computer disks, which offer high storage capacity combined with very rapid access.

The system is computer-driven by two floppy disks; one controls the system and the other contains the editing information and cues. Time code can be used to synchronise to tape or video recorders which it will follow instantly. It can therefore act as the equivalent of up to eight extra tracks containing inserts which can have been programmed from a 'file' of prerecorded material instantly available in any order of sequence. An important facility is the ability to adjust the sync of the inserts backwards or forwards, which can be of considerable value in relation to sound effects or language dubbing.

228

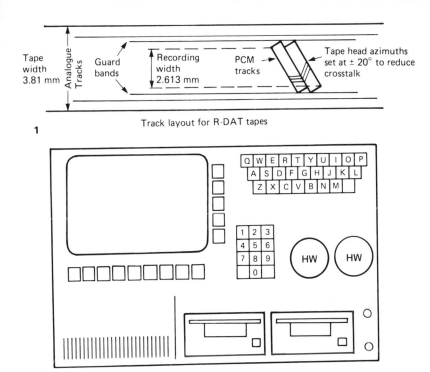

1 Track layout for R-DAT tapes

1. *Track layout for R-DAT tapes.* The combination of low-mass transport, light tape and tension, together with very sophisticated servo control, makes the R-DAT recorder particularly suitable for rapid selection of material and editing. It is possible to rewind the whole of the tape in less than 15 s and park it within 1/48 000 s of the required cue, where it can be put into record or replay in 0.1 s. It thus becomes possible to programme edits, even in random order, or make very quick jump cuts between pre-programmed cues.

2. *The Audiofile.* Controls for the Audiofile are arranged around a monitor screen and consist of 'soft' keys, the functions of which are displayed on the monitor adjacent to each key and change according to the system in use at the time. There is also a typewriter-keyboard for typing in cue titles and instructing the computer, and a numeric keypad to key in time-code information. For accurate editing the programme roughly in the area of the edit point is dumped into a memory, access to which is controlled by a hand-wheel (HW). By turning the hand-wheel the recording can be rocked backwards and forwards to locate the exact edit point.

S-DAT

There is also a static-head DAT (S-DAT) system, which uses a magnetoresistive head for playback. The magnetic flux on the tape is not used to generate the signal but to modulate it. As this is not dependent upon the rate of change of the flux the tape/head speed is not so critical and a static head can be used. These machines are still in the development stage.

Exposure to loud sounds for lengthy periods can result in permanent deafness.

The Dangers of High Sound Levels

No book about sound recording would be complete without mention of the dangers of exposure to excessive loudness levels. Medical evidence has confirmed that long exposure to very loud sound can cause damage to the hearing mechanism, resulting in permanent deafness.

The range of human hearing

Our ears encounter an enormous range of audio power. It could vary from about $0.000\,000\,001$ W for a soft whisper to $10\,000$ W for a turbo-jet engine (i.e. $30-170$ dB relative to 10^{-12} W).

To cope with this range, the human ear has a built-in limiting mechanism. When subjected to a loud sound our ears switch in a sort of attenuator so that we become partially deaf.

If the sound is not too loud or the exposure too long, our hearing recovers soon after the sound has stopped. The louder the sound and the longer the exposure to it, the longer it will take for our hearing to recover. This could take a few hours or days. In the extreme case, where exposure to loud sound has been long and not interspersed with long periods of rest, physical damage to the cochlea will result in permanent deafness.

One of the dangers of exposure to loud sounds is that the resulting deafness can come on gradually and therefore not be noticed until the condition is acute and the situation irreversible. Impairment usually starts with a dip in the 4000 Hz region which widens and deepens until there is a general loss of intelligibility.

There is a definite relationship between loudness, duration of exposure and the effect on hearing. Although it varies between individuals, the relationship between duration of sound and hearing impairment is generally linear but the effect of loudness rises as loudness increases and above 100 dB it goes up very rapidly indeed.

Whereas hitherto exposure to very loud sounds was rare and of short duration (except for those engaged in noisy industrial processes), nowadays with the availability and popularity of high power amplifiers particularly in the 'pop' scene it is possible to listen to sound levels of over 100 dB for long periods of time.

Sound levels in discotheques often exceed 120 dB and there is evidence to suggest that many young people have permanent hearing impairment as a result.

Recording engineers engaged in recording pop music, whose clients have already damaged their own hearing and therefore demand higher and higher levels, are particularly at risk. If it is your profession, your livelihood could be at stake.

SOUND POWER AND POWER LEVEL SCALES

Power (W)	Power level (dB re 10^{-12} W)	Source
25–40 million	195	Saturn rocket
100 000	170	Ram jet
10 000	160	Turbo-jet engine with afterburner
1 000	150	4 propeller airliner
100	140	75 piece orchestra ⎧ peak RMS
10	130	Pipe organ ⎨ levels in 0.125 s
1	120	Piano ⎩ intervals
0.1	110	Blaring radio
0.01	100	Car on motorway
0.001	90	Voice shouting (average long time RMS)
0.000 1	80	
0.000 01	70	Voice, conversational level (average long time RMS)
0.000 001	60	
0.000 000 1	50	
0.000 000 01	40	
0.000 000 001	30	Voice, very soft whisper

VALUES OF SOUND PRESSURE LEVEL IN SPECIFIC FREQUENCY BANDS WHICH INDICATE A HAZARD TO HEARING AT STATED DAILY DURATIONS FOR ONE EXPOSURE

Octave band specified as centre frequency (Hz)	Sound pressure levels at specified durations (dB)					
	4 hours	2 hours	1 hour	30 min	15 min	7 min
63	100	103	106	110	116	122
125	94	97	100	104	110	116
250	90	93	96	100	106	112
500	87	90	93	97	103	109
1000	85	88	91	95	101	107
2000	83	86	89	93	99	105
4000	82	84	88	92	98	104
8000	81	84	87	91	97	103

In Conclusion

To sum up we can define the main functions of the sound element in television as:

1. To provide much of the substance of the programme in terms of information and entertainment value, in an artistic and technically satisfactory manner.

2. To increase the realism and effectiveness of the pictures by simulating the appropriate acoustic conditions to suit the apparent environment and by creating the illusion of solidity of the settings.

3. To suggest movement and the third dimension by matching sound perspective to the picture viewpoint. In television particularly, owing to the small size of the screen, distance tends to be depicted in depth rather than in width. The ability to convey a sense of distance and relative perspective is an important property of sound. If the sound perspective matches the picture it will underline the realism. If it disagrees it will undermine it.

4. To enhance the mood and impact of the production with suitable music and sound effects.

The sound operation

The process of sound operations involves selecting, controlling and blending sound sources, i.e. the exercise of choice on behalf of the listener. In television this must be achieved with the minimum of restriction to the visual presentation, which involves considerable flexibility of approach. No single system of sound pickup can apply to all circumstances. Each successive problem must be considered separately and the most suitable method applied individually. Full use should be made, as necessary, of sound processing techniques such as frequency response control, automatic volume compression and the addition of artificial reverberation, etc. Effective, creative sound requires imaginative forethought at the planning stage, based upon a thorough knowledge of the problems and possibilities of the medium, followed by judgment in selecting and controlling the right equipment.

The effect of the picture

The presence of a picture does not detract from the importance of the sound, or make it less critical; it merely presents an extra dimension and a greater challenge.

It is worth taking care.

Glossary

Absorption coefficient (24) The ratio of acoustic energy absorbed by a surface exposed to a sound field to that incident on the surface. It is equal to 1 minus the reflection coefficient of the material.

Acoustic effect (22) The effect of the surrounding environment on the sound.

Acoustic treatment (28) The application of a material to the surface of an enclosure to modify the acoustic.

Ambient sound (22, 64) The noise, reverberation or atmospheric sounds that form a background to the principal source.

Amplitude The peak value of a waveform.

Attack time (104) The time taken for a gain reduction to take effect in a compressor or limiter.

Audience reinforcement (PA) (128) The relaying of programme sound to the audience present at a production. The term 'reinforcement' as opposed to PA (public address) suggests that the live sound would be audible to the audience, albeit at low volume, without the use of reinforcement, but this is seldom the case with present microphone technique.

Aural perspective (138) The impression of distance of a sound source created by such factors as relative volume, proportion of direct to reverberant sound, characteristic quality etc.

Azimuth (222) The angle between the gap of a tape head and the longitudinal axis of the tape.

Backing (sound) (192) Accompaniment.

Backing (vision) (138) Background scenery.

Backing track Pre-recorded accompaniment to which a vocal or solo part is added.

Balance (20) The selection of microphone positions and volume levels from a number of sources to produce an artistically correct effect.

Bandwidth (14) The interval between cut-off frequencies.

Beats The periodic variations of amplitude resulting from the addition of two slightly different frequencies.

Bel (16) A unit used to compare the magnitude of powers. It is equal to the logarithm to the base 10 of the ratio of the powers; one bel = 10 decibels.

Bit (214) Contraction of 'binary digit' (a 1 or a 0); 8 bits make a byte.

Boom (78) A mobile device incorporating a telescopic arm for varying the position and direction of a microphone to follow mobile action.

Capacitor A component made up of conducting plates separated by insulation (termed dielectric).

Capacitor microphone (42) Also called *condenser microphone*. A microphone in which the capacitance is varied by sound causing movement in one plate (diaphragm) in relation to the fixed backplate.

Cassette (222) A reel-to-reel tape system in which both the feed and the take-up spools are incorporated in a plastic box.

CCIR A recording standard (Comité Consultatif International des Radiocommunications).

Clean feed (136) An output feed to a contributor which contains all the contributions except his own.

Clean feed talkback A circuit which normally carries clean feed but can be intercepted by a talkback microphone and switch.

Closed circuit A programme not intended for transmission to the public at the time.

Coincident pair (72) An arrangement of microphones for stereophony in which two microphones are placed with their diaphragms pointing in different directions but so close together (possibly in the same case) that the path length between the sound source and the microphones is virtually the same.

Coloration (14) A form of distortion of a sound signal caused by the addition of spurious harmonics.

Companding (204) A method of applying volume compression to a signal prior to processing (e.g. recording) and a complementary expansion afterwards, resulting in a reduction in system noise in the final output.

Composite sound Complete programme sound, i.e. commentary with international sound (effects, music etc. normally available to all language commentary areas).

Compressor (104) A device for reducing automatically the dynamic range of a signal.

Condenser microphone See *capacitor microphone.*

Conference network (136) Circuits connecting a number of remote points to provide simultaneous speech to or from any points (remotes).

Crab (82) To move sideways.

Cue programme A circuit carrying audio programme material, used for cueing or monitoring purposes.

Cycle (14) One complete sequence of a variation which occurs in a periodic manner.

DAC (214) Digital to analog converter.

DASH recorder (218) Digital Audio Static Head recorder.

dBA When describing the loudness of a sound in relation to its sound pressure level it is necessary to take into account the unequal sensitivity of our ears to sounds of different frequencies and intensities (page 16). Sound pressure level is, therefore, usually measured through a 'weighting' network which roughly corresponds to the equal loudness contour. Such measurements are termed A weighted and the units are dBA.

DBX (210) A noise reduction system which employs pre-emphasised compression and de-emphasised expansion.

Dead acoustics (24) An area where there is very little reverberation.

Decibel (16) A measure of relative intensity used for acoustic measurements because it obeys a logarithmic law, as does the human ear in relating sound intensity to sensation. The decibel is one tenth of a bel. It represents about the minimum change of level that can be appreciated by the ear, and that only on steady tone.

De-emphasis A response that decreases with frequency.

Diffraction (10) The manner in which sounds are able to bend around obstacles with dimensions smaller than the wavelength of the sound.

Diffusion (24) The distribution of reflective paths for sound waves within an enclosure.

Digital sound (214) A process of sound processing in which the normal analog waveform of an audio signal is converted to a series of numerical measurements which can be described by a digital code. Usually a binary code is used and the processing equipment only has to recognise two alternative conditions, 1 (on) or 0 (off), so the system can be made very robust and distortion-free.

Dolby A noise reduction process named after its inventor, Dr Ray Dolby. Dolby A (204) is a frequency-selective companding arrangement intended for professional use, mainly for the production of master tapes. Dolby B (206) and Dolby C (208) are processes applied to commercial recordings, particularly cassettes, on the understanding that there will be a corresponding process applied in the reproducing equipment.

Dolby SR Spectral Recording process. The action is similar to Dolby C, except that three levels of action staggering are used: high-level, mid-level, and low-level. A main signal path conveys the high-level signals. Side-chain signals are additively combined with the main signal in the encoding mode and subtractively in the decoding mode. The intention is to keep all low-level signal components fully boosted at all times. When a high-level component at a particular frequency occurs it is cut back, without affecting the low-level components of different frequency, and restored in the decoding process.

Drop-in (218) The process of inserting recorded audio by playing up to a chosen point and switching from playback to record mode.

Dropout (216) Momentary loss of signal due to fault in tape coating or dust.

Dubbing (212) The process of re-recording from one recording to another. The term is also used for the addition of sound, e.g. dialogue or effects, to a previously recorded picture.

Ducking (118) A method of using a compressor in which the volume of one signal is controlled by another.

Dynamic range (16, 231) The ratio (expressed in phons) between the softest and loudest sound. Or the ratio (in decibels) of signal variation available between overload and unacceptable signal/noise level.

Echo (22) Discrete, separately identifiable repetitions of sound due to reflections from hard surfaces. The term is often used, wrongly, to mean reverberation (which is due to multiple overlapping reflections).

Eigentones (115) Standing-wave resonances set up in an enclosure when the reflective path lengths correspond to the wavelength of a sound.

Electret diaphragm (42) A diaphragm that has been given a permanent electrostatic charge, thereby eliminating the need for a polarising voltage with condenser microphones. Alternatively the charge can be applied to the backplate of a condenser microphone.

Electrostatic microphone (42) Also called *condenser microphone*, it operates by the variations in capacitance between the diaphragm and the backplate spaced closely behind it. The capacitive reactance is connected in series with a high resistance across which the output voltage is developed.

Equalizer (110) A response-shaping filter.

Erasure (198) The process of removing previously recorded signals from magnetic materials prior to recording.

Expander (104) An amplifier that increases in gain as amplitude increases, the opposite of compressor.

Eye line The direction in which an artist looks while performing.

Feedback A proportion of the output of an amplifying system that is returned to the input.

Fishpole (74) A short pole supporting a microphone carried by an operator, used to pick up mobile sources of sound in cramped conditions or outside locations where a boom would be inappropriate.

Flanging (114) A phasing effect caused by making two outputs of the same source and then recombining them in varying phase.

Flat (30) A piece of hard scenery.

Frequency modulation (222) A method of modulating a carrier by varying its frequency as opposed to its amplitude.

Frequency (14) The rate of repetition of a periodic function measured in cycles per second or hertz (Hz).

Fuzz (194) Deliberately introduced distortion for special effect with electric guitars.

Gain (16) Amplification of an amplifier or component, usually expressed in decibels.

Graphic equaliser (110) A frequency response shaping filter in which the spectrum is usually divided into octave or third-octave bands. Control is by sliders, the positions of which gives a 'graphic' indication of the shape of the response curve.

Guide track (196) A specially recorded sound track (normally one track of a multi-track recording) used for synchronising purposes. If, for example, an orchestral accompaniment is to be added live to a pre-recorded vocal, the guide track, possibly containing a bar count and metronome lead-in, can be fed to the conductor on headphones to keep the orchestra in time during an intro or non-vocal reprise.

Haas effect (116) The effect that determines the apparent direction of a sound source. When the same sound is reproduced simultaneously from two or more places it will appear to come from the nearest one. The relationship between sound volume, time delay and directionality is known as the Haas effect.

Harmonic (14) A sinusoidal oscillation having a frequency which is an integral multiple of the fundamental frequency. It is the harmonics (or in musical terms upper partials) that shape the waveforms and make it possible to distinguish between various instruments, even when they are playing the same note.

Harmonic distortion (18) The production of spurious harmonics due to a distortion of the original waveform.

Hertz (Hz) (14) Unit of frequency: one hertz equals one cycle of repetition per second.

Howlround (128) Instability in a loop, usually including an electrical and acoustic path, e.g. loudspeaker and microphone, where the electrical amplification is gerater than the acoustic loss. If the gain is progressively increased the initial result is an uneven increase in response due to positive feedback peaking at the frequency at which the electro/acoustic circuit is most sensitive, finally building up into a continuous howl at this frequency.

Indirect sound (24) Sound which reaches the listener or microphone by acoustic reflection.

Intensity of sound (16) The intensity is a measure of the power of a sound. It is usually measured in decibels relative to the threshold of hearing at 1000 Hz. One dB represents roughly the smallest perceptible change of sound intensity. It is not the same as loudness, measured in *phons* (which relate to the intensity at 1000 Hz), because the ear has an unequal frequency response, which also varies with intensity.

Intermodulation distortion (18) A form of distortion caused by one component frequency-modulating another, thereby creating spurious sum and differences which produce a 'rough' tone.

International sound (136) Feed of clean effects (without commentary) to be fed to several countries for addition to commentaries in their own language.

Lazy arm (44) A simple form of boom or counterweighted arm mounted on a stand that can be used for suspending a microphone over an object such as a music stand or piano.

Leader (200) A section of uncoated tape, usually coloured white, which is joined to the beginning of a recording to allow for threading to the take-up spool. It sometimes also carries identification.

Level (100) The intensity of steady tone used for test purposes and for lining-up equipment. Zero level is normally taken as a power of 1 mW in a resistance of 600 ohms.

Limiter (104) A device for preventing the volume of a signal from rising above a pre-set value, thereby preventing overload. The action is similar to a compressor except that the gain reduction is more severe. A compression ratio of 10:1 or more can be considered as limiting, as a very large increase in input is required to make a significant difference to the output.

Loudness (16) The subjective impression of the strength of a sound. Loudness is affected by a number of factors such as the actual volume of the sound, the listener's aural sensitivity, the masking effect of one source of sound on another of similar frequency and 'irritation factor' (the listener's appreciation of the material). Unwanted sound (noise) tends to sound louder than wanted sound of similar volume.

Masking (204) The manner in which the ability to hear sounds of a particular frequency becomes reduced in the presence of louder sounds of similar frequency.

Mega Prefix meaning one million (M).

Micro Prefix meaning one millionth part (μ).

Monophonic (68) The reproduction of sound via a single medium of transmission (mono).

Mute (film) Silent film.

Mute (instrument) (180) Device for reducing the tone and altering the characteristic quality of a musical instrument.

Obstacle effect (38) Sound waves are able to curve around obstacles whose dimensions are smaller than their wavelength. This effect also depends on the shape, i.e. the streamlining of the object.

Optical sound (162) A film with a sound track which takes the form of variation in light transmission. It is reproduced by shining a light source through the optical track on to a photo-electric cell.

Oversampling (216) When converting between digital and analog the sampling rate must be at least twice the highest frequency it is intended to convert (see Nyquist limit). If the sampling rate is just high enough to satisfy Nyquist it is necessary to employd a very steep ('brick wall') low-pass filter to prevent the formation of heterodyne frequencies which cause spurious noises and distortion. Unfortunately, very steep filters introduce phasing distortion, so oversampling can be used to effectively increase the sampling rate (e.g. by a factor of four) so that a much more gentle filter can be employed with consequently less distortion.

Nyquist limit (214–216) In analog to digital conversion (ADC) and DAC the sampling rate must be at least twice the highest frequency it is intended to convert. This is called the Nyquist limit. If it is exceeded a form of heterodyning will occur, due to the sidebands overlapping the base band, and the reconstruction filter in the DAC will not be able to separate them. The result is *aliasing*, a difference frequency caused by beating between the input and the sampling rate. In practice the sampling rate has to be at least 2.2 times the highest audio frequency. The extra 0.2 is to allow for filtering out the out-of-band audio.

PA See *audience reinforcement*.

Pan To change direction in the lateral sense.

Pan-pot (panoramic potentiometer) (70) Control for adjusting the apparent position of sound in a stereo image.

Parametric equalizer (110) A frequency shaping device which allows both the frequency and the bandwidth of the boost or cut to be selected.

Peak programme meter (100) An instrument for measuring programme sound in terms of peaks averaged over a specified period. It has a rapid rise time and a slow recovery, so it is easy to read on programme.

Phantom power (42) Method of sending DC supply to a condenser microphone by connecting the positive supply to both signal wires and the negative to the (earthed) screen.

Phase (14) The position in the cycle that a waveform has reached at any given instant. Waves are said to be in phase when their position in the cycle coincides.

Phon (16) A measure of the loudness of a sound that takes into account the unequal frequency response of our ears. Phons and decibels are the same at 1000 Hz. At other frequencies their relationship is illustrated by the curves of equal loudness.

Pitch (16) Subjective effect of sound, related mainly to frequency but affected also by intensity and harmonic structure. As volume is increased high sounds can seem higher and low sounds lower.

Post sync (196) The addition of dialogue, music or effects to synchronise with a previously recorded picture.

Pre-delay (113) The application of delay to the input of an artificial reverberation device to simulate the time taken for the reflections to build up in the real situation.

Pre-hear (106) A circuit, normally controlled by a key or button, by which it is possible to listen to the input of a fader or control before it is faded up.

Presence filter (110) A filter which imparts a rise, usually in the form of a peak to the frequency response in the region of 3–7 kHz, which is the frequency range in which the sibilants in vocal sounds tend to lie. This improves clarity of diction and gives the effect of closeness (i.e. presence) to the sound.

Pressure gradient operation (34) Method of operation in which both sides of a microphone diaphragm are exposed to the sound waves. The signal is generated by the difference in pressure (pressure-gradient) existing at each moment in time due to the phase change between front and back.

Pressure operation (34) Method of operation in which the diaphragm is only exposed to the sound waves on one side.

Proximity effect (56) Increase in low frequency response which occurs when a point source (e.g. mouth) is close to a pressure-gradient microphone.

Pulse code modulation (214) A system in which the audio signal is converted into a series of pulses. The signal information can be contained in the position of the pulses relative to a marker pulse (pulse position modulation), the width of the pulses, i.e. mark/space ratio, or their amplitude. The latter is generally used in digital audio. Using binary notation all above 50% can be given the value 1 and all below 0. So the system is very robust.

Recovery time (104) The time taken for a compressor or limiter to return the gain to normal after the removal of a high-level signal.

Reverberation (22) The sustaining effect of multiple reflections inside an enclosure.

Reverberation time (22) The time taken for a sound to die away through 60 dB (one-millionth of its original intensity) in a reverberant enclosure after it has been cut off.

Roll-back and mix (196) The process of repeating, by means of a recording, the end of a previous sequence in order for it to be mixed into the start of the next sequence.

Sampling (214) The process of taking samples of an analog signal at intervals in time for converting to digital.

Sound image (26) The mental picture created by a particular character of sound.

Stereophony (66) A two-channel audio system which provides the illusion of spatial distribution of the sources in the horizontal plane.

Talkback (94) Open microphone to one or more destinations. Used for cueing and briefing instructions.

Unidirectional microphone A microphone that picks up sound from one direction only.

VCA (108) Voltage-controlled amplifier. The gain of a VCA is controlled by a small DC voltage (typically −2 to +10 V) applied via a fader in the control desk.

VU meter (100) A meter for indicating programme volume which indicates signal power (in decibels) on steady tone, and volume units (percentage utilisation of the channel) on programme.

Wavelength (14) The distance between corresponding parts of a waveform.

White noise A full-spectrum signal having the same energy level at all frequencies.

Wild track (196) Sound recording to accompany a picture but not synchronised to it.

Windscreen (62) A device fitted around a microphone to protect it from wind or turbulence due to rapid movement through the air, e.g. on a boom. Windscreens have to be large to be effective against wind. Smaller close-talking (pop) screens are available for some microphones to guard against breath puffs.

Winchester disk (228) A magnetic disc with a very high storage capacity used for storing digital information.

Wow A slow variation of pitch, normally most noticeable on low notes in recordings, due to speed instability in the recording or reproducing mechanism.

Zero level The level used for reference and for lining up equipment using standard tone. It corresponds to a power of 1 mW in 600 ohms and is indicated by the figure 4 on a peak programme meter.